1st PAGES Young Scientists Meeting – Retrospective Views on Our Planet's Future 2009

AA009055

IOP Conference Series: Earth and Environmental Science
Volume 9

Corvallis, Oregon, USA
6 - 7 July 2009

Editors:

Caroline Cleroux
Jennifer Fehrenbacher
Steven Phipps
Summer Rupper
Branwen Williams
Thomas Kiefer

ISBN: 978-1-5108-0202-5
ISSN: 1755-1307

Printed from e-media with permission by:

Curran Associates, Inc.
57 Morehouse Lane
Red Hook, NY 12571

Some format issues inherent in the e-media version may also appear in this print version.

Printed by Curran Associates, Inc. (2015)

For permission requests, please contact the Institute of Physics at the address below.

Institute of Physics
Dirac House, Temple Back
Bristol BS1 6BE UK

Phone: 44 1 17 929 7481
Fax: 44 1 17 920 0979

techtracking@iop.org

Additional copies of this publication are available from:

Curran Associates, Inc.
57 Morehouse Lane
Red Hook, NY 12571 USA
Phone: 845-758-0400
Fax: 845-758-2634
Email: curran@proceedings.com
Web: www.proceedings.com

Table of contents

Volume 9

PAGES 1ST Young Scientists Meeting (YSM) – 'Retrospective Views on our planet's future' 6–7 July 2009, Corvallis, USA

Editors: Caroline Cléroux, Jennifer Fehrenbacher, Steven Phipps, Summer Rupper, Branwen Williams and Thorsten Kiefer

Accepted papers received: 10 March 2010
Published online: 12 April 2010

Reconstructing and Understanding Climate Forcings and Feedbacks (PAGES Focus 1)

Regional and Global-scale Earth System Dynamics (PAGES Foci 2 and 3)

Past Human-Climate-Ecosystem Interactions (PAGES Focus 4)

Improving and Understanding Paleoscientific Methods (PAGES Cross-Cutting Themes 1-3)

IOP Conf. Series: Earth and Environmental Science **9** (2010) 012001 doi:10.1088/1755-1315/9/1/012001

Response of terrestrial N_2O and NO_x emissions to abrupt climate change

M Pfeiffer and J O Kaplan

Ecole Polytechnique Fédérale de Lausanne, ARVE Group, Station 2,
1015 Lausanne, Switzerland

E-mail: mirjam.pfeiffer@epfl.ch

Abstract. Being a potent greenhouse gas, N_2O emitted by the terrestrial biosphere during abrupt climate change events could have amplified externally forced warming. To investigate this possibility, we tested the sensitivity of terrestrial N_2O emissions to an abrupt warming event by applying the ARVE-DGVM in combination with a novel scheme for process-based simulation of terrestrial N_2O and NO_x emissions at the Gerzensee site in Switzerland. In this study, we aim to quantify the magnitude of change in emissions for the abrupt climate change event that occurred at the transition from Oldest Dryas to Bølling during the last deglaciation. Using high-resolution multiproxy records obtained from the Gerzensee that cover the Late Glacial, we apply a prescribed vegetation change derived from the pollen record and temperature and precipitation reconstructions derived from $\delta^{18}O$ in lake sediments. Changes in soil temperature and moisture are simulated by the ARVE-DGVM using the reconstructed paleoclimate as a driver. Our results show a pronounced increase in mean annual N_2O and NO_x emissions for the transition (by factor 2.55 and 1.97, respectively), with highest amounts generally being emitted during summer. Our findings suggest that summertime emissions are limited by soil moisture, while temperature controls emissions during winter. For the time between 14670 and 14620 cal. years BP, our simulated N_2O emissions show increase rates as high as 1% per year, indicating that local reactions of emissions to changing climate could have been considerably faster than the atmospheric concentration changes observed in polar ice.

1. Introduction

To understand the role of the terrestrial biosphere in influencing climate change and air quality, it has become increasingly important to quantify the magnitude and variability of trace gas emissions from the earth's land surface to the atmosphere, e.g., [1]. The ice core record of atmospheric N_2O concentrations shows considerable temporal variability that closely parallels polar temperature records [2], [3], [4], [5], indicating that nitrogen trace gas emissions may be very sensitive to climate change. Variability in atmospheric N_2O and NO_x concentrations is linked to nitrogen cycling in the oceans, e.g., [6] and the terrestrial biosphere [1]. Given the high greenhouse warming potential of N_2O [7] it is conceivable that increases in its atmospheric concentration could have amplified ongoing climate change during periods of rapid warming [8].

Over the last decade, a number of studies have attempted to quantify N_2O emissions both for the present-day and the past based on measurement-based inventories of N_2O concentrations in ice cores [2], [3], [4], [5]. In addition, several research groups have attempted to model regional and global source emissions of N_2O and NO_x, e.g [9], [10], [11]. However, process-based, mechanistic modeling of nitrogen trace gas emissions using Dynamic Global Vegetation Models (DGVMs) is still at an early stage, and only a few attempts have been made so far to successfully apply a process-oriented nitrogen cycle model in a DGVM [12], [13], [14], [15]. The implementation of a novel scheme for simulating emissions of N_2O and NO_x into the ARVE-DGVM - part of the current study - will allow us to explicitly simulate the emissions of different nitrogen trace gas species (e.g., N_2O, NO_x) on regional to global scales.

From Greenland ice cores it is known that the transition from Oldest Dryas (~18000 - 14650 cal. yrs. BP) to Bølling (14650 - 14000 cal. yrs. BP). was linked to a significant increase in atmospheric N_2O concentration from approx. 210 ppbv to 260 ppbv [2]. Analyses of lake sediment and pollen records at the Gerzensee site in Switzerland have provided reconstructions of both mean annual air temperature based on $\delta^{18}O$ of lake carbonates and other proxies, e.g., [16], [17], and changes in vegetation cover from pollen records, e.g. [18]. The very similar oxygen isotope records from Greenland ice cores and European lake sediments during the Glacial Termination [19] suggest that drastic climatic changes occurred quasi-simultaneously on hemispheric scale [16]. Greenland summit is reported to have warmed 9 ±3 °C [20], while the temperature of the warmest month in Western central Europe increased between 3 °C [21] and 5.5 °C [22] within approx. one century. We hypothesize that the pronounced and abrupt change in climate conditions caused an increase in N_2O and NO_x emissions at the Gerzensee site. Our aim for the study presented here is to quantify the proportional difference of N_2O and NO_x emissions before and after the abrupt climate warming at the Gerzensee site in Switzerland that occurred at the end of the last glacial.

2. Materials and methods

Gerzensee (46.830 °N, 7.547 °E) is located on the Swiss Plateau approx. 15 km southeast of Bern, at 602 m above sea level. The lake was formed as a kettle-hole after the last glaciation, and has a surface area of 0.27 km^2, a catchment area of 2.6 km^2, and a maximum depth of 10 m. A number of studies performed at the Gerzensee in recent years investigate how abrupt climate change effects ecosystem behavior on a local to regional scale using different proxies such as pollen, plant macrofossils, $\delta^{18}O$, chironomids and charcoal, e.g., [23], [24], [25], [26], [27], [28], [29].

For the time of Termination 1, quantitative July temperature reconstructions using fossil chironomid and pollen assemblages, as well as bulk sediment oxygen isotopes from Gerzensee have been carried out for the period between 15000 and 13000 cal. years BP [17]. The excellent correlation between the Gerzensee bulk sediment $\delta^{18}O$ and the layer-counted North GRIP ice core provides a precise, high-resolution chronology for the record.

To simulate the impact of the rapid warming on emissions of N_2O and NO_x, we used the $\delta^{18}O$ derived July temperature anomaly reconstruction based on isotopic analysis of shell carbonate picked from the lake sediments (*U. von Grafenstein et al.* pers. comm.) to infer monthly mean temperatures during the deglaciation from current monthly mean temperatures at Gerzensee. The original $\delta^{18}O$ derived temperature anomalies for the period 15350cal. years BP to 14000 cal. years BP have roughly decadal resolution, and we used linear interpolation in order to obtain annual temperature anomaly

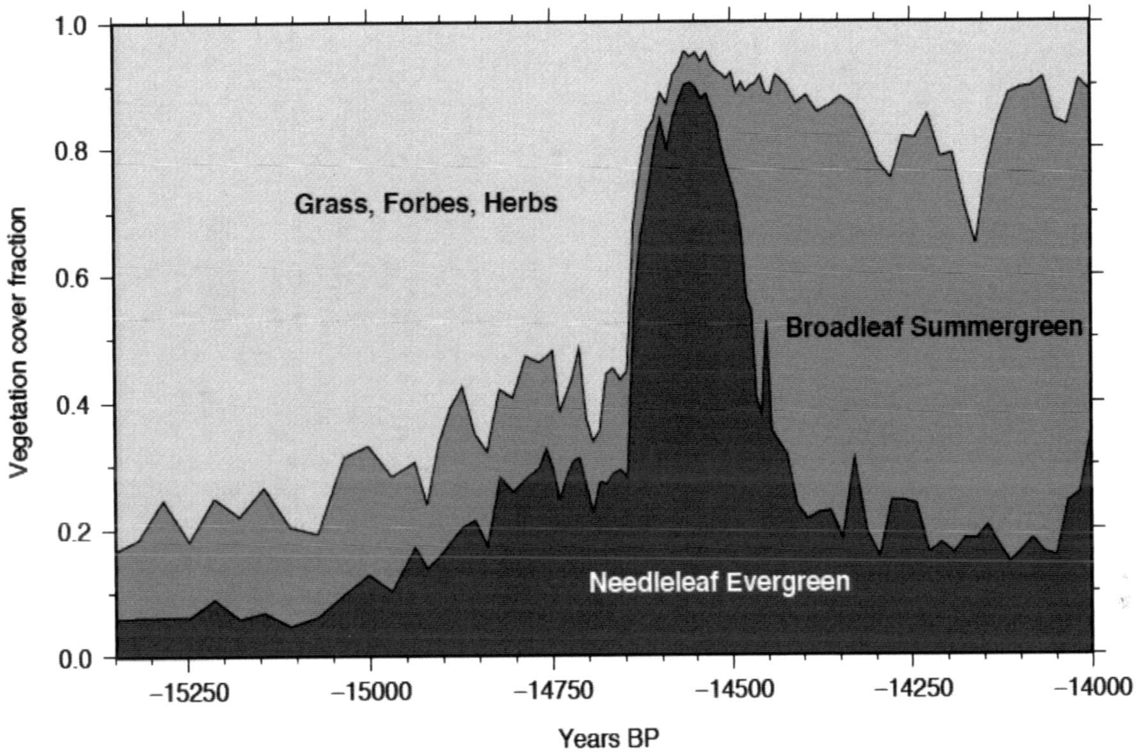

Figure 1: Vegetation cover at Gerzensee during the transition from Oldest Dryas to Bølling, reconstructed from pollen accumulation rates [29].

values between years. We used the WGEN-based weather generator [27] implemented into ARVE-DGVM to create daily values for temperature and precipitation from monthly means, and ran its soil physics submodel to simulate daily soil volumetric water content and soil temperature. We therefore did not run the ARVE-DGVM as a full DGVM for this study. Daily precipitation was generated based on the pollen-based reconstruction of precipitation at Lac Lautrey [22] (130 km from Gerzensee, 46.587 °N, 5.864 °E) for the Termination 1 transition. Annual precipitation reconstructions for Lac Lautrey before and after the transition were scaled to present-day annual precipitation at Lac Lautrey. The corresponding scaling factor was applied to present day annual precipitation at Gerzensee to obtain an estimate for annual precipitation amounts before and after the transition at Gerzensee.

The daily soil temperature and soil moisture values derived from ARVE-DGVM are required as drivers for the nitrogen compartment model, which includes routines to calculate litterfall and soil organic matter pools and was adapted from the ORCHIDEE-CN nitrogen module ([15], *S. Zaehle*, pers. comm.). The nitrogen model was parameterized by prescribing typical values published in literature for C/N ratios, biomass and soil organic matter for the three plant functional types (PFTs) that were present at Gerzensee during the transition (grass/herbal, needleleaf evergreen, broadleef summergreen, figure 1). Given the current model setup, the nitrogen budget is not closed and vegetation changes are not modelled directly. N_2O and NO_x emissions are simulated separately for each PFT. Total emission values per unit area were obtained in a post-processing step by scaling the amount emitted by each PFT according to the prescribed fractional covers of the PFTs estimated from pollen accumulation rates from Gerzensee for the time of the transition (*W.O. van der Knaap*, pers. comm., [31]).

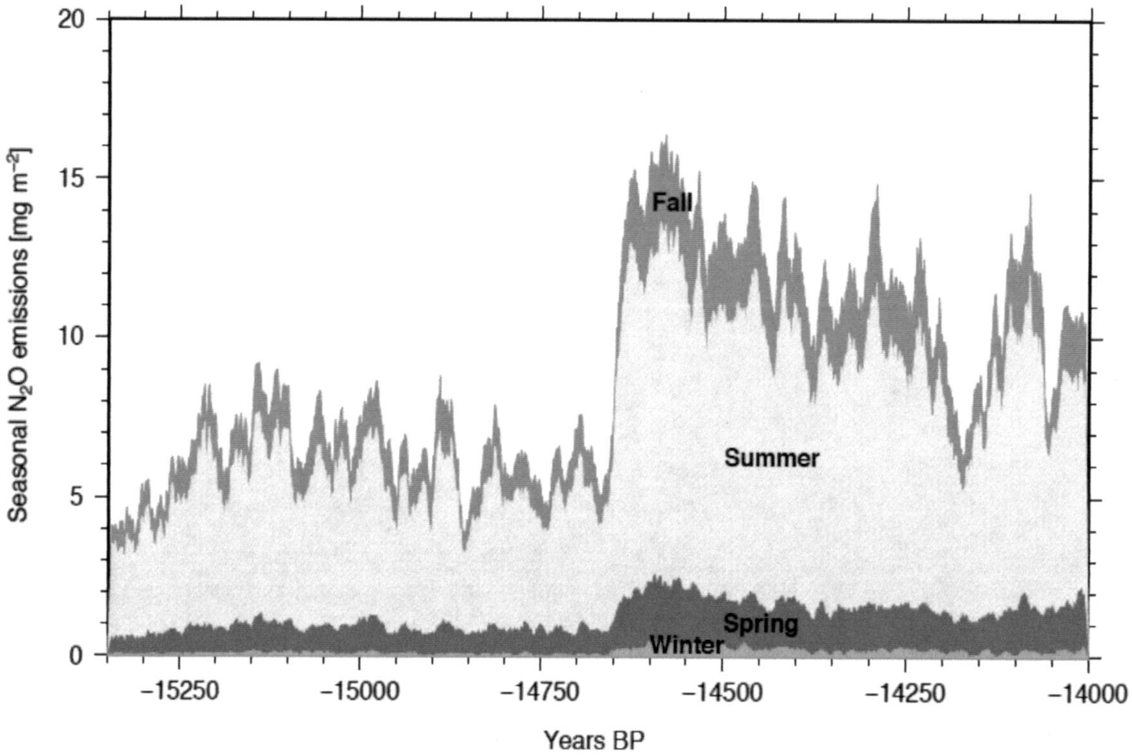

Figure 2: Simulated N_2O emissions for the Oldest Dryas to Bølling transition at Gerzensee (Switzerland). Shown are total annual emissions (top margin of curve) and the breakdown of total emissions into seasonal fractions.

Table 1: Statistics of seasonal and annual N_2O and NO_x fluxes before and after the transition (SD: Standard Deviation; SEM: Standard Error in the Mean)

	N_2O [mg m^{-2} timespan^{-1}]					NO_x [mg m^{-2} timespan^{-1}]				
	spring	summer	fall	winter	annual	spring	summer	fall	winter	annual
	before transition (14750 BP - 14660 BP)					*before transition (14750 BP - 14660 BP)*				
Median	0.713	3.920	0.908	0.004	5.831	13.189	56.963	14.731	0.082	87.931
Mean value	0.759	4.031	1.030	0.043	5.863	13.318	58.580	16.061	0.755	88.715
1 SD	0.360	1.645	0.555	0.093	1.839	3.614	15.263	5.304	1.531	17.624
SEM	0.038	0.173	0.059	0.010	0.194	0.381	1.609	0.559	0.161	1.858
Minimum value	0.192	1.203	0.275	0.000	2.195	6.712	29.665	8.452	0.007	53.940
Maximum value	1.718	11.292	3.411	0.517	13.650	23.600	125.291	38.583	7.678	160.517
	after transition (14630 BP - 14550 BP)					*after transition (14630 BP - 14550 BP)*				
Median	1.799	10.092	2.386	0.218	14.568	23.248	113.944	30.125	3.024	171.153
Mean value	1.813	10.336	2.467	0.356	14.972	23.431	116.219	30.804	4.621	175.074
1 SD	0.539	2.809	0.795	0.362	3.210	4.924	24.222	7.291	4.215	28.542
SEM	0.060	0.310	0.088	0.040	0.354	0.544	2.675	0.805	0.465	3.152
Minimum value	0.792	4.690	0.995	0.009	8.147	14.789	68.180	17.550	0.195	108.531
Maximum value	3.168	16.732	4.637	1.702	22.128	37.139	178.884	51.762	18.742	238.429

3. Results

The combined increase in temperature and precipitation during the Oldest Dryas/Bølling transition resulted in a pronounced increase in simulated N_2O (figure 2, table 1) and NO_x emissions for the Gerzensee site. As we ran our simulations for just one single site, we cannot directly compare the magnitudes of the simulated changes with the ice core record, but the direction of the simulated pattern is consistent with the increase in atmospheric N_2O concentration observed in Greenland ice [2]. For NO_x, mean annual emissions increased by factor 1.97 from 88.7 (SD ±17.6; SEM ±1.86) mg m^{-2} yr^{-1} to 175.1 (SD ±28.5; SEM ±3.15) mg m^{-2} yr^{-1} when comparing the average of the time period 14750 – 14660 BP (before the transition) to the average of the time period 14630 – 14550 BP (after the transition). For N_2O, the increase was even more pronounced with a factor of 2.55 from 5.9 (SD ±1.8; SEM ±0.19) mg m^{-2} yr^{-1} to 15.0 (SD ±3.2; SEM ±0.35) mg m^{-2} yr^{-1}. Similar variability characterizes the seasonal amplitude of N_2O emissions at present day, e.g., a factor of 1.82 for the spruce site at Höglwald/Germany in 1995 (derived from figure 4 in [32]).

The annual mean soil moisture averaged over the 90 years prior to the transition increased by factor 1.64 from 16.80 (SD ±2.52; SEM ±0.27) % to 27.62 (SD ±2.72; SEM ±0.30) % as compared to the annual mean soil moisture averaged over the 80 years after the transition. The simulated difference in annual mean soil temperature between the coldest year within the 90 years before the transition and the warmest year of the 80 years after the transition was 4.9 °C, as compared to 6 °C for $\delta^{18}O$ derived July air temperature. The increase in soil temperature is less pronounced than the increase in air temperature due to the dampening effect in soils. Moreover, the simulated 4.9 °C increase refers to annual mean temperature, while the proxy-derived 6 °C increase refers to the increase of July temperature only. Annual mean soil temperature averaged over the 90 years before the transition increased by 2.5 °C from 5.6 (SD ±0.5; SEM ±0.06) °C to 8.1 (SD ± 0.5; SEM ±0.06) °C as compared to the annual mean soil temperature averaged over the following 80 years after the warming. As there is variability within both parts of the temperature time series, comparison of the averages over both time series shows a less pronounced increase in temperature than a comparison of extreme years before and after the transition.

The seasonal distribution of N_2O emissions shows a clear maximum for summer (JJA) and a minimum for winter (DJF). For all seasons, the change from generally cold and dry conditions to warmer and moister conditions leads to an increase in emissions when comparing the 90 years before the transition to the 80 years after the warming. For spring and summer, mean N_2O emissions increased by factor 2.39 from 0.8 (SD ±0.4; SEM ±0.04) mg m^{-2} to 1.8 (SD ±0.5; SEM ±0.06) mg m^{-2} and by factor 2.56 from 4.0 (SD ±1.6; SEM ±0.17) mg m^{-2} to 10.3 (SD ±2.8; SEM ±0.31) mg m^{-2}, respectively. Fall emissions increased by factor 2.40 from 1.0 (SD ±0.6; SEM ±0.06) mg m^{-2} to 2.4 (SD ±0.8; SEM ±0.09) mg m^{-2}. Winter emissions increased by factor 8.20, but in general were extremely low and showed a very high variability (< 0.1 mg m^{-2} before the transition, and 0.4 (SD ±0.4, SEM ±0.04) mg m^{-2} after the transition). The seasonal pattern and general behavior of NO_x emissions (not shown here) are very similar to that of N_2O, with an increase of emissions by factor 1.76, 1.98, 1.92 and 6.12 for spring, summer, fall and winter, respectively.

4. Discussion

Nitrification (aerobic oxidation of NH_4^+ to NO_3^-), and, even more importantly, denitrification (anaerobic reduction of NO_3^- to N_2), are the main processes of the terrestrial nitrogen cycle that result in N_2O and NO_x emissions from soils. Both processes are catalyzed by microorganisms whose kinetics is temperature dependent. Soil moisture exhibits an important control on denitrification, which takes place where anaerobic conditions exist in water-filled pores of soil. Apart from soil temperature and soil moisture, a group of other factors such as substrate quantity (depending on biomass and litter production), nitrogen input (deposition, biological N-fixation), and litter quality (C/N ratio of litter, which influences degradability) influence production and release of N_2O from soils. The interplay of all these factors makes it hard to foresee which factors are dominantly influencing N_2O emissions. Initial sensitivity tests with our nitrogen model for the Oldest Dryas/Bølling transition at Gerzensee reveal that annual emissions of both trace gases show a good correlation with mean annual temperature (R^2 = 0.81) and mean annual soil moisture (R^2 = 0.75) when both factors are tested independently. This indicates that temperature and moisture are important factors influencing emissions. In addition to temperature and precipitation, the observed increase of emissions is reinforced by the change in vegetation cover from mostly herbaceous tundra vegetation to mostly forest (figure 1). A comparison of our vegetation change scenario to a sensitivity test where vegetation cover was kept constant with the conditions of the first simulation year for the entire simulation period reveals that approx. 33% of the increase in emissions is due to the change in vegetation cover. In our simulations, grasslands tend to produce less N_2O than forests, therefore the transition from mostly grassland to mostly forest cover amplifies the increase in emissions caused by the changes in meteorology. However, the phenomenon of pristine temperate grasslands producing less nitrogen trace gas emissions than forests under comparable climate conditions requires further evaluation.

As shown in figure 2, the magnitude of N_2O emissions during summer increases along with the variability of emissions. Before the transition, summer soil temperatures were high enough to allow microbial processes to result in the net emissions of some nitrogen trace gases. The temperature and moisture increases linked to the abrupt climate change event stimulates microbial activity and therefore emissions, but at the same time, the limiting effect of soil moisture during drier years becomes more pronounced. Total summer emissions are generally higher, which results in increased variability of emissions between drier years and wetter years. In contrast, wintertime emissions are controlled mainly by temperature rather than soil moisture. The increase in winter soil temperature leads to a decrease of days with soil frost and thus promotes microbial activity. Based on these model results, we suggest that emissions in summer are likely to be limited by the availability of water in the soil, while temperature controls emissions during winter.

With this study we could show the effect of soil temperature and soil moisture changes on N_2O and NO_x emissions. We did not intend to consider all aspects in the nitrogen cycle affecting nitrogen availability and nitrogen trace gas emissions yet, which requires simulation of a closed nitrogen cycle with actual vegetation dynamics instead of using prescribed biomass values from literature. A closed nitrogen budget with direct simulation of vegetation dynamics using the full DGVM with the nitrogen cycle directly integrated into it is a goal for future applications and will allow to address all effects of environmental changes on nitrogen availability, such as changes in quality and quantity of plant biomass, litter production, soil organic matter and microbial activity, as well as succession processes. Moreover, we only can judge the relative change in emissions at the moment, as the simulated total quantities of trace gas emissions per time unit still need to be evaluated. We used high-resolution reconstructions of mean annual temperature and precipitation for the Oldest Dryas/Bølling transition as a basis for creating a daily weather scenario to drive the nitrogen model. With roughly decadal resolution, the $\delta^{18}O$ proxy record of the Gerzensee is an exceptionally detailed record of the abrupt climate change event. However, we acknowledge that our method of interpolating this paleoproxy record to an annual timestep, and then simulating daily weather based on these data with a weather generator, introduces additional uncertainty into our results. Nevertheless, we feel that the resulting data used to drive our model represents a plausible scenario of the climate changes that occurred during the Bølling warming event.

5. Conclusions

Our simulated N_2O emissions for the Oldest Dryas/Bølling transition at Gerzensee are in good accordance with the timing of the concentration changes observed in Greenland ice. We were able to show that terrestrial N_2O and NO_x emissions react very sensitively to small changes in temperature, moisture and vegetation cover, and can change significantly within a very short time span. As we only simulate emission changes for a single site, we cannot directly transfer our results into estimates of atmospheric concentration changes. For the time between14670 and 14620 cal. years BP, our simulated N_2O emissions show increase rates as high as 1% per year. This indicates that short-term local reactions of emissions to changing climate could have been considerably more pronounced than visible from the atmospheric concentration changes recorded in Greenland ice (concentration change from approx. 210 to 258 ppmv from 14800 to 14500 cal. years BP, annual increase rate of approx. 0.06%). In the future we will expand the focus of our research from single sites such as Gerzensee to global simulations, allowing more direct comparison with ice core data. Moreover, additional simulations at Gerzensee with more detailed reconstructions of palaeoclimatic conditions, in particular a better way of reconstructing precipitation, will provide more information about the reaction of nitrogen trace gas emissions from terrestrial ecosystems to abrupt climate change.

Acknowledgements
We would like to acknowledge S. Zaehle for sharing source code from ORCHIDEE-CN, as well as B. Ammann, A. F. Lotter, U. von Grafenstein and W.O. van der Knaap for sharing unpublished data material. Funding for this work was provided by grants from the Swiss National Science Foundation (PP0022_119049) and the Italian Ministry for Research and Education (FIRB RBID08LNFJ) for the Research Project CASTANEA.

References

[1] Overpeck J, Whitlock C and Huntley B 2003: Terrestrial biosphere dynamics in the climate system: past and future; In: *Paleoclimate, Global Change and the Future*, ed K D Alverson *et al* (Springer) pp 81- 111
[2] Flückiger J, Dällenbach A, Blunier T, Stauffer B., Stocker T F, Raynaud D and Barnola J-M 1999: Variations in atmospheric N_2O concentration during abrupt climatic changes; *Science* **285** 227-30
[3] Sowers T, Alley R B and Jubenville J 2003: Ice core records of atmospheric N_2O covering the last 106.000 years; *Science* **301** 945-48
[4] Spahni R, Chappelaz J, Stocker T F, Loulergue L, Hausammann G, Kawamura K, Flückiger J, Schwander J, Raynaud D, Masson-Delmotte V and Jouzel J 2005: Atmospheric methane and nitrous oxide of the Late Pleistocene from Antarctic ice cores; *Science* **310** 1317- 21
[5] Wolff E and Spahni R 2007: Methane and nitrous oxide in the ice core record; *Philosophical Transactions of the Royal Society* A **365** 1775-92
[6] Suthof A, Ittekkot V and Gaye-Haake B 2002: Millenial-scale oscillations of denitrification intensity in the Arabian Sea during the late Quaternary and its potential inflence on atmospheric N_2O and global climate; *Global Biogeochem. Cycles* **15** 637-49
[7] Forster P, Ramaswamy V, Artaxo P, Berntsen T, Betts R, Fahey D W, Haywood J, Lean J, Lowe D C, Myhre G, Nganga J, Prinn R, Raga G, Schultz M and van Dorland R 2007: Changes in atmospheric constituents and radiative forcing, in: *Climate change 2007: The Physical Science Basis. Contribution of Working Group I to the Fourth Assessment Report of the Intergovernmental Panel on Climate Change*, ed S Solomon *et al* (Cambridge, UK and New York, NY, USA: Cambridge University Press,)
[8] Woodwell G M, Mackenzie F T and Houghton R A, Apps M, Gorham E and Davidson E 1998: Biotic feedbacks in the warming of the Earth; *Climatic Change* **40** 495-518
[9] Potter C S and Klooster S A 1998: Interannual variability in soil trace gas (CO_2, N_2O, NO) fluxes and analysis of controllers on regional to global scales, *Global Biogeochem. Cy.* **12**, 621-35
[10] Parton W J, Holland E A, Del Grosso S J, Hartman M D, Martin R E, Mosier A R, Ojima D S and Schimel D S 2001: Generalized model for NO_x and N_2O emissions from soils, J.

Geophys. Res.-Atmospheres **106** 17403-19

[11] Sorai M, Yoshida N and Ishikawa M 2007: Biogeochemical simulation of nitrous oxide cycle based on the major nitrogen processes, *J. Geophys. Res.* **112**, G01006, doi:10.1029/2005JG000109

[12] Sokolov A P, Kicklighter D W, Melillo J M, Felzer B S, Schlosser C A and Cronin T W 2008: Consequences of considering carbon-nitrogen interactions on the feedbacks between climate and the terrestrial carbon cycle, *J. Climate* **21**, 3776-96

[13] Xu-Ri and Prentice I C 2008: Terrestrial nitrogen cycle simulation with a dynamic global vegetation model, *Glob. Change Biol.* **14** 1745-64

[14] Thornton P E, Doney S C, Lindsay K, Moore J K, Mahowald N, Randerson J T, Fung I, Lamarque J-F, Feddema J J and Lee Y-H 2009: Carbon-nitrogen interactions regulate climate-carbon cycle feedbacks: results from an atmosphere-ocean general circulation model, *Biogeosciences Discuss.* **6** 3303-54

[15] Zaehle S and Friend A D 2009: Carbon-nitrogen cycle dynamics in the ORCHIDEE-CN land surface model: Part 1: Model evaluation and sensitivity, *Global Biogeochem. Cy., accepted,* **2009**

[16] Schwander J, Eicher U and Ammann B 2000: Oxygen isotopes of lake marl at Gerzensee and Leysin (Switzerland), covering the Younger Dryas and two minor oscillations, and their correlation to the GRIP ice core; *Palaeogeography, Palaeoclimatology, Palaeoecology* **159** 203-14

[17] Lotter A F, Heiri O, Brooks S, van Leeuwen J N F, Eicher U, Schwander J and Ammann B. 2008: Rapid summer temperature changes during Termination 1: multi-proxy reconstructions from Gerzensee (Switzerland); *Geophysical Research Abstracts* **10**, EGU2008-A-09661

[18] Lotter A F, Birks H J B, Eicher U, Hofmann W, Schwander J and Wick L 2000: Younger Dryas and Allerød summer temperatures at Gerzensee (Switzerland) inferred from fossil pollen and cladoceran assemblages; *Palaeogeography, Palaeoclimatology, Palaeoecology* **159** 349-61

[19] von Grafenstein U, Eicher U, Erlenkeuser H, Ruch P, Schwander J and Ammann B. 2000: Isotope signature of the Younger Dryas and two minor oscillations at Gerzensee (Switzerland): palaeoclimatic and palaeolimnologic interpretation based on bulk and biogenic carbonates; *Palaeogeography, Palaeoclimatology, Palaeoecology* **159** 215-29

[20] Severinghaus J P and Brook E J 1999: Abrupt climate change at the end of the last glacial period inferred from trapped air in polar ice; *Science* **29** 930-34

[21] Heiri O and Millet L 2005: Reconstruction of Late Glacial summer temperatures from chironomid assemblages in Lac Lautrey (Jura, France); *Journal of Quaternary Science* **20/1** 33-44

[22] Peyron O, Bégeot C, Brewer S, Heiri O, Magny M, Millet L, Ruffaldi P, Van Campo E and Yu G 2005: Late-Glacial climatic changes in Eastern France (Lake Lautrey) from pollen, lake-levels and chironomids; *Quaternary Research* **64** 197-211

[23] Ammann B, Birks H J B, Brooks S J, Eicher U, von Grafenstein U, Hofmann W, Lemdahl G, Schwander J, Tobolski K and Wick L. 2000: Quantification of biotic responses to rapid climatic changes around the Younger Dryas — a synthesis; *Palaeogeography, Palaeoclimatology, Palaeoecology* **159** 313-47

[24] Ammann B 2000: Biotic responses to rapid climatic changes: Introduction to a multidisciplinary study of the Younger Dryas and minor oscillations on an altitudinal transect in the Swiss Alps; *Palaeogeography, Palaeoclimatology, Palaeoecology* **159** 191 - 201

[25] Brooks S J 2000: Late-glacial fossil midge stratigraphies (Insecta: Diptera: Chironomidae) from the Swiss Alps, *Palaeogeography, Palaeoclimatology, Palaeoecology* **159** 261-79

[26] Hofmann W 2000: Response of the chydorid faunas to rapid climatic changes in four alpine lakes at different altitudes; *Palaeogeography, Palaeoclimatology, Palaeoecology* **159** 281-92

[27] Lemdahl G 2000: Lateglacial and Early Holocene insect assemblages from sites at different

altitudes in the Swiss Alps—implications on climate and environment; *Palaeoclimatology, Palaeoecology* **159** 293-312

[28] Tobolsky K and Ammann B 2000: Macrofossils as records of plant responses to rapid Late Glacial climatic changes at three sites in the Swiss Alps, *Palaeogeography, Palaeoclimatology, Palaeoecology* **159** 251-59

[29] Wick L 2000: Vegetational response to climatic changes recorded in Swiss Late Glacial lake sediments; *Palaeogeography, Palaeoclimatology, Palaeoecology* **159** 231-50

[30] Richardson C W and Wright D A 1984: WGEN: A model for generating daily weather variables; United States Department of Agriculture, Agriculture Research Service, ARS-8, August 1984

[31] van der Knaap W O 2009: Estimating pollen diversity from pollen accumulation rates: a method to assess taxonomic richness in the landscape; *The Holocene* **19** 1 159-63

[32] Stange F, Butterbach-Bahl K, Papen H, Zechmeister-Boltenstern S, Li C and Aber J 2000: A process-oriented model of N_2O and NO emissions from forest soils, 2. Sensitivity analysis and validation; *J. Geophys. Res.* **105**, D4, 4385-98

Evaluating source, chemistry and climate change based upon the isotopic composition of nitrate in ice cores

Meredith G Hastings[1]

Brown University, Department of Geological Sciences and Environmental Change Initiative, 324 Brook Street, Box 1846, Providence, RI, USA 02912

Email: Meredith_Hastings@brown.edu

Abstract. A clear negative trend is found in the nitrogen isotopic ratios ($^{15}N/^{14}N$) of nitrate over the industrial period, based on a 100-meter ice core from Summit, Greenland. This record indicates that isotopic composition of ice core nitrate reflects changes in nitrogen oxide (NO_x) source emissions and that post-industrial emissions of NO_x have resulted in a 12‰ (vs. air N_2) decline in $\delta^{15}N$ of atmospheric nitrate over the last ~300 years. Interestingly, over the last glacial period (as recorded in the Greenland GISP2 ice core), the $\delta^{15}N$ of nitrate changes by ~20‰ from a typical pre-industrial Holocene value of 10‰ to a mean glacial value of 28‰, despite the lack of a significant change in nitrate concentration. The more recent ice core record clearly indicates an influence of NO_x emission sources, therefore suggesting that the glacial-interglacial change in $\delta^{15}N$ may be a record of significant variations in the contribution of NO_x sources, such as lightning, biomass burning, biogenic soil emissions and oxidation of nitrous oxide in the stratosphere. In contrast to the source changes recorded by the nitrogen isotopes, the oxygen isotopic record of atmospheric nitrate ($^{18}O/^{16}O$, $^{17}O/^{16}O$) in ice cores has implications for reconstruction of past atmospheric oxidant levels. This is because the oxygen isotopic composition of nitrate reflects oxidation of NO_x to nitrate in the atmosphere by oxidants such as ozone and hydroxy and peroxy radicals. Measurements of variations in the isotopic composition of atmospheric nitrate inferred from ice cores could be used to improve interpretation of other records of $\delta^{15}N$ such as those from tree rings and/or sediments.

1. Introduction and background

The formation of nitric acid (HNO_3), or nitrate (NO_3^-), in the atmosphere is the main sink of reactive nitrogen oxides (NO_x). NO_x is an important chemical constituent in the global atmosphere because of its reactivity and its role in regulating the oxidizing capacity of the atmosphere through its connection to ozone and hydroxyl radical. The sources of NO_x include both anthropogenic and natural sources, with a majority of the global burden currently contributed by fossil fuel burning and the rest primarily contributed by emissions from biomass burning, nitrification/denitrification in soils, lightning, and oxidation of nitrous oxide in the stratosphere.

The lifetime of tropospheric NO_x against conversion to HNO_3 in the modern atmosphere is on the order of less than a day to about a week depending upon latitude and altitude [1]. Nitric acid is very soluble and is readily lost from the atmosphere via wet (e.g. precipitation) and dry (e.g. dust particles) deposition. The relatively short lifetime of NO_x (and HNO_3) and the temporal and spatial variability of

[1] To whom correspondence should be addressed.

emissions sources result in a very non-uniform distribution of NO_x, making it difficult to understand the variability of NO_x with measurements alone. As a result, atmospheric chemical transport models are employed to investigate NO_x/NO_3^- cycling, and are often used to understand the influence of NO_x on the oxidizing efficiency of the atmosphere, which in turn regulates the lifetimes of many other reactive traces gases in our atmosphere including methane. Many of these models test the accuracy of their representation of NO_x/NO_3^- cycling by comparing with measurements of NO_3^- in wet and dry deposition around the world [2], although most observations are from North America and Europe. Simulated and observed NO_3^- deposition is the result of a complex mixture of natural and anthropogenic source emissions that vary in space and time, plus the transport of those emissions, chemical transformation of NO_x to NO_3^-, and finally, deposition. All of these variables contribute to the lifetime of NO_x and the ultimate concentration of NO_3^- that is deposited.

The isotopic composition of atmospheric NO_3^- represents a tool that can be used to further our understanding of NO_x/NO_3^- cycling, as well as give us the ability to diagnose atmospheric models beyond concentration alone. Recent methodological advances have facilitated a new era of research into the isotopic composition of atmospheric NO_3^- and its utility as tracer of NO_x sources and chemistry. It is now clear that the oxygen isotopic composition of NO_3^- ($\delta^{18}O$, $\Delta^{17}O$)[2] formed in the atmosphere is a tracer of the oxidants that convert NO_x to NO_3^- [3,4]. Observations of atmospheric NO_3^- in the modern atmosphere typically range from 40 – 85‰ in $\delta^{18}O$ and 20 – 35‰ in $\Delta^{17}O$. Ozone (O_3) has a uniquely high $\delta^{18}O$ and $\Delta^{17}O$ ($\delta^{18}O \sim 100$‰, $\Delta^{17}O \sim 35$‰) compared to other oxidants in the atmosphere such as H_2O, OH or O_2 ($\delta^{18}O$ ranging from ~ -40 to +30‰, $\Delta^{17}O \sim 0$‰). Simplistically, the range in observations can be thought of as a result of mixing between NO_3^- formed via reaction of NO_2 and OH that results in a relatively low $\delta^{18}O$ (and low $\Delta^{17}O$), versus NO_3^- formed from reactions of NO_2, O_3, and a compound containing H, which results in a relatively high $\delta^{18}O$ (and high $\Delta^{17}O$) representative of the influence of ozone. On the other hand, the nitrogen isotopic composition of NO_3^- is representative of NO_x source signatures, chemical transformation processes in the atmosphere, and/or physical processing of NO_3^- in the atmosphere or post-depositionally [4,5,6]. The $\delta^{15}N$ of NO_3^- in the modern atmosphere has been shown to range from -15 to +15‰ [7]. It has been difficult to prove a direct connection between $\delta^{15}N$ of NO_3^- and NO_x sources for several reasons: many records represent total nitrogen, not atmospheric nitrate deposition alone; after the nitrate is deposited it can undergo biological processing or physical and/or chemical changes that can alter the original signal; and/or the isotopic signatures of atmospheric nitrate sources are not well quantified. What is significant about the findings presented here is that it is the first archived record that clearly shows a change in the isotopic composition of nitrate that cannot be explained except by a significant change in the source of atmospheric nitrate.

2. The nitrogen isotopic composition of nitrate in Greenland ice

The recent measurement technique advances have contributed the ability to measure the isotopic composition of NO_3^- in ice core samples [8,9,10], which tend to have very low concentrations of NO_3^- (e.g. <5 μmol l^{-1}) and are limited in sample volume. The data shown herein utilizes the denitrifier method for the isotopic determination of nitrate in ice core samples from Greenland. This method uses denitrifying bacteria to convert as little as 10 nanomoles of nitrate in a water sample to nitrous oxide gas, which is then measured using an isotope ratio mass spectrometer in continuous flow mode [8,9].

Figure 1 below shows the results of NO_3^- concentration and $\delta^{15}N$ measurements from ice core samples collected at Summit, Greenland (72.5°N, 38.4°W, 3200 meters above sea level). Samples from two different ice cores are shown: the more recent record (the last ~300 years) comes from a 100-meter ice core drilled at Summit in 2006 [11], while the glacial/interglacial record (>5000 years

---

[2] $\delta^{18}O$, in units of ‰ vs. VSMOW, is defined as $\delta^{18}O = (R_{sample}/R_{standard} - 1) \times 1000$‰ where $R = {}^{18}O/{}^{16}O$; for $\delta^{15}N$, $R = {}^{15}N/{}^{14}N$ and the standard is air N_2. In most of the work cited here, $\Delta^{17}O$ ("capital delta O-17") is approximated as $\Delta^{17}O \approx \delta^{17}O - \delta^{18}O \times 0.52$.

before present) represents ice samples from the deep ice core drilled by the Greenland Ice Sheet Project (GISP2) in 1993.

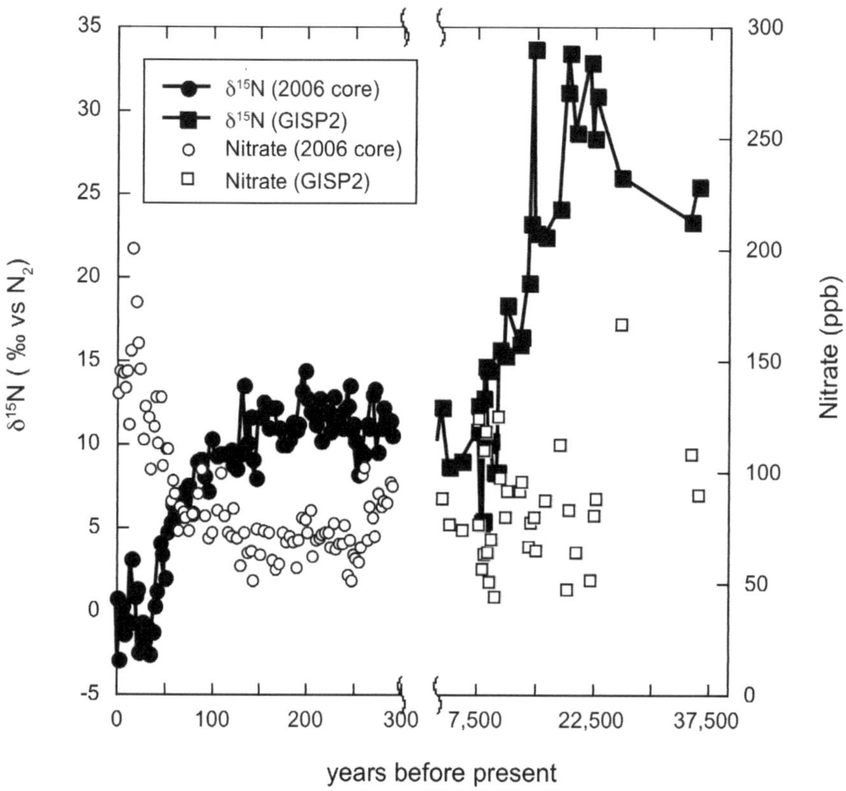

Figure 1. Nitrate concentration (open symbols) and $\delta^{15}N$ of nitrate (closed symbols) in Greenland ice over the last ~36,000 years. Data from two cores are shown: a 100-meter core drilled in 2006 and samples from the GISP2 deep ice core drilled in 1993. Note the break in the scale on the x-axis.

The $\delta^{15}N$ of nitrate in Greenland ice over the last 300 years shows a clear negative trend that follows the introduction of nitrogen oxides (NO_x) from fossil fuel combustion starting ~1850 AD (Figure 1). As also seen in other cores from Greenland, the concentration of nitrate shows a clear increase beginning in the early decades of the 20th century. Interestingly, the isotopes of nitrate change significantly throughout the two records with small or no change in the concentration of nitrate in ice. It is likely then that the isotopes of nitrate are a much more sensitive indicator of changes in the environment than nitrate concentration.

The industrial period shows a change in $\delta^{15}N$ of ~12‰ between today and 300 years ago, while a change of ~20‰ is observed between the pre-industrial Holocene (typically ~9‰) and a mean glacial value of ~30‰. Several hypotheses were discussed in [12] to try to explain the large change observed across the glacial/interglacial transition. The glacial environment was very different than that during the Holocene: the glacial atmosphere was much drier; accumulation decreased by about half and primarily occurred during the summer at Summit compared to the Holocene; nitrate was scavenged by dust during the glacial period compared to a greater tendency for wet deposition of nitric acid today; presumably there were significant changes in oxidation chemistry and the budget of NO_x. In the last 300 years, these changes have either not occurred or have not been nearly as extreme as over the glacial/interglacial transition. What has changed significantly is the introduction of a new source of NO_x, namely fossil fuels combustion. Furthermore, none of the changes in the atmosphere described above would be expected to follow a trend of the type that is observed in the recent $\delta^{15}N$ record (Figure 1). We thus draw the conclusion that the more recent ice core record clearly indicates an influence of NO_x emission sources. Is it possible then that the glacial-interglacial change in $\delta^{15}N$ is a record of significant variations in the contribution of NO_x sources, such as lightning, biomass burning, nitrous oxide oxidation and/or biogenic soil emissions?

Physical processing does not appear to explain the glacial/interglacial change that occurs in $\delta^{15}N$ of nitrate. The changes in accumulation – less accumulation in the glacial and primarily occurring in the summer – raise concern that post-depositional processing of nitrate in surface snow might be more likely. Several recent studies have investigated the role of post-depositional processing in determining the isotopic signals of nitrate preserved in ice [13,14,15,16]. It is clear that at low accumulation sites, such as Dome C on the Antarctic plateau, a great deal of nitrate loss occurs resulting in enrichment of ^{15}N relative to ^{14}N in nitrate left in the surface snow [14,16]. At relatively high accumulation sites, such as Summit, study of the snow and atmosphere just above the snow in the summer suggest that some loss of nitrate occurs (as a result of photolysis of nitrate), but most appears to be recycled back to the snow resulting in minimal to no change in the $\delta^{15}N$ of nitrate in surface snow. Thus we do not expect that post-depositional processing as a result of the change in accumulation during the glacial can explain the change in $\delta^{15}N$. Further, a comparison of isotopes of nitrate in GISP2 ice (Summit) and Dye 3, a Greenland site that today experiences twice as much accumulation as Summit, do not show any difference in $\delta^{15}N$. (Still, it should be noted that a threshold might exist between low (Dome C, ~3 g cm^{-2} yr^{-1}) and high accumulation (Summit, ~20 g cm^{-2} yr^{-1}) at which post-depositional processing is minimized. This should be explored further to determine whether it is reasonable to conclude that post-depositional loss of nitrate was not significant at Summit during the glacial period). Another potential effect to consider during the glacial period is that HNO_3 was primarily scavenged by alkaline dust particles that contained Ca^{2+}, compared to more recent time periods when HNO_3 is primarily taken up by water/ice surfaces. We expect the unidirectional uptake of nitrate onto dust, producing $Ca(NO_3)_2$, to result in a kinetic isotope effect with isotopically lighter particulate NO_3^- than gas-phase HNO_3. This scenario produces the opposite of the change observed between interglacial and glacial ice.

Returning to the question posed above, can the observed changes in $\delta^{15}N$ be explained by changes in the sources of NO_x? The current state of knowledge of the isotopic signatures associated with NO_x sources is weak. Table 1 details the published values for $\delta^{15}N$ of NO_x from direct measurements (unless otherwise indicated). This collection suggests that the production of NO_x in the stratosphere from the oxidation of nitrous oxide is the only source that results in a high, positive $\delta^{15}N$. All of the pre-industrial measurements of $\delta^{15}N$ of nitrate in Greenland ice are positive in $\delta^{15}N$ and values greater than 22‰ are observed (Figure 1, Table 1). The significant decrease in $\delta^{15}N$ following the introduction of fossil fuels combustion on Northern Hemisphere continents could suggest that this source produces NO_x with a considerably negative $\delta^{15}N$. Another hypothesis is that the balance of NO_x source contributions over the last ~300 years has changed such that a source which produces NO_x with a positive $\delta^{15}N$ decreased substantially, while the addition of fossil fuel combustion, with a small negative source signature, has increased substantially. The isotopic signatures of NO_x emitted from 'natural' soils, biomass burning, and North American fossil fuels combustion are currently unknown. There is a clear need to better quantify and/or quantify for the first time the isotopic signatures associated with these NO_x sources. This will give us the ability to tease out the influence of different NO_x sources on the nitrate preserved in ice cores from sites with sufficient accumulation. This represents an exciting possibility – a quantitative tracer of the pre-industrial sources of NO_x represents a connection between the biosphere (biomass burning, soils), atmosphere (lightning, chemistry, transport), and climate. Additionally, this tracer could be used to diagnose and test atmospheric models that include the sources and chemistry of nitrogen oxides and their influence on the oxidizing capacity of the atmosphere. Finally, the relatively short lifetime of NO_x and nitrate means that regional scale changes over time can be interpreted based on ice cores from different locales.

Table 1. Published values for $\delta^{15}N$ of NOx emission sources.

Source of NO_x	$\delta^{15}N$ of NO_x (‰ vs N_2)	Geographic location of study	Study material	Reference
Vehicle emissions	+3.4 to +3.9	Colorado, USA	Tailpipe exhaust	Moore, 1977
Vehicle emissions	-13 to +2	Pretoria, South Africa	Tailpipe exhaust	Heaton, 1990
Vehicle emissions	+5.7	Swiss Middle Land, Solothurn, Switzerland	Roadside denuder collections	Amman et al., 1997
Coal combustion	+6 to +10	Pretoria, South Africa	Smokestacks	Heaton, 1990
Soils	-48.9 to -19.9[a]	Guangzhou, China	Soil samples fertilized in laboratory	Li and Wang, 2008
Lightning	-0.5 to +1.4		Passing air through electric sparks in laboratory	Hoering, 1957
N_2O oxidation in stratosphere	+19±3		Calculated based on known composition and fractionation of stratospheric N_2O	Savarino et al., 2007
Snowpack photodenitrification	~≥ -15[b]			Morin et al., 2008
Biomass burning	?[c]			

[a]depends upon $\delta^{15}N$ of inorganic N (i.e., nitrate, ammonium) in soils
[b]depends upon $\delta^{15}N$ of nitrate (substrate) in snow
[c] $\delta^{15}N$ of *total nitrogen* on aerosol particles collected in biomass burning plumes range from +10.6 to +25.7‰

3. The utility of the isotopes of atmospheric nitrate in ice

The ability to measure the isotopic composition of nitrate at high resolution in ice greatly advances the knowledge that can be gained from nitrate in ice cores. At high accumulation sites it appears that the isotopic composition of nitrate is representative of that in the atmosphere. This means that it is possible that this tracer will hold quantitative information regarding sources of NO_x and how they change over time. (An additional note is that the oxygen isotopic composition of nitrate holds information about the chemical oxidation pathways that produce nitrate in the atmosphere and this additional knowledge will influence our understanding of past concentrations of ozone and hydroxyl radicals and their influence on the composition of the atmosphere [17]. The $\delta^{18}O$ of NO_3^- does not show a significant change in the last 300 years, though the last glacial period appears to be significantly higher in $\delta^{18}O$ than the pre-industrial Holocene). At low accumulation sites, where nitrate is subject to significant post-depositional processing, the isotopic composition of nitrate may not directly represent the atmospheric loading, but this tracer is still proving useful. Recent measurements at Dome C show that the $\delta^{15}N$ of nitrate archived in ice varies with the exposure of nitrate in surface snow to radiation, thus potentially serving as a recorder of solar activity over time [16]. It will be important in future studies to quantify how the accumulation of snow influences what information is preserved in the isotopes of nitrate in ice, particularly beyond what we now consider very low accumulation (Dome C, ~3 g cm^{-2} yr^{-1}) and relatively high accumulation (Summit, ~20 g cm^{-2} yr^{-1}).

A significant implication of the more recent ice core record is that it should be possible to use the nitrogen isotopic composition of nitrate to quantify the sources that contribute to nitrate deposition. In the present environment this represents an opportunity to trace sources of acid deposition and air pollution; reconstructed records over time represent the possibility of linking changes in biospheric and atmospheric sources of NO_x and changes in climate. Ice core records of nitrate reflect atmospheric nitrate deposition, and the records shown here could be then used to elucidate additional information from the $\delta^{15}N$ of total N found in tree rings and ocean sediments.

Acknowledgements
The data presented in this manuscript was supported by the National Science Foundation Office of Polar Programs (0454803) and the Joint Institute for Study of the Atmosphere and Ocean (under NOAA Cooperative Agreement No. NA17RJ1232). Participation in the PAGES 2009 YSM and OSM meetings was sponsored by PAGES through a grant from the U.S. National Science Foundation and by the Brown University ADVANCE program.

References
[1] Levy II H, Moxim WJ, Klonecki AA, and Kasibhatla PS 2001 *J. Geophys. Res.* **104** 26,279-306
[2] Lamarque J-F, *et al* 2007 *J. Geophys. Res.* **110** D19303
[3] Michalski G, Scott Z, Kabiling M, Thiemens MH 2003 *Geophys. Res. Lett.* **30** 1870
[4] Hastings MG, Sigman DM, Lipschultz F 2003 *J. Geophys. Res.*, **108(D24)** 4790
[5] Freyer HD, Kley D, Volz-Thomas A, Kobel K 1993 *J. Geophys. Res.* **98** 14791-6
[6] Morin S, *et al* 2008 Science **322** 730
[7] Kendall C, Elliott EM, Wankel SD 2007 *Stable Isotopes in Ecology and Environmental Science (2nd edition)* ed R Michener and K Lajtha (Hoboken, NJ: Blackwell) chapter 12 pp 375-449
[8] Sigman DM, Casciotti KL, Andreani M, Barford C, Galanter M, Böhlke JK 2001 *Anal. Chem.* **73** 4145-53
[9] Casciotti KL, Sigman DM, Hastings MG, Böhlke JK, Hilkert A 2002 *Anal. Chem.* **74** 4905-12
[10] Kaiser J, Hastings MG, Houlton BZ, Röckmann T, Sigman DM 2007 *Anal. Chem.* **79** 599-607
[11] Hastings MG, Jarvis JC, Steig EJ 2009 *Science* **324** 1288
[12] Hastings MG, Sigman DM, Steig EJ 2005 *Glob. Biogeochem. Cyc.* **19** GB4024
[13] Hastings MG, Steig EJ, Sigman DM 2004 *J. Geophys. Res.* **109** D20306
[14] Blunier T, Floch GL, Jacobi H-W, Quansah E 2005 *Geophys. Res. Lett.* **32** L13501
[15] Jarvis JC, Hastings MG, Steig EJ, Kunasek S 2009 *J. Geophys. Res.* **114** D17301
[16] Frey MM, Savarino J, Morin S, Erbland J, Martins JMF 2009 *Atmos. Chem. Phys.* **9** 8681-96
[17] Alexander B, Hastings MG, Allman DJ, Dachs J, Thornton JA, Kunasek S 2009 *Atmos. Chem. Phys.* **9** 5043-56

IOP Conf. Series: Earth and Environmental Science **9** (2010) 012003 doi:10.1088/1755-1315/9/1/012003

Tropical vegetation evidence for rapid sea level changes associated with Heinrich Events

Catalina González[1,2], Lydie M. Dupont[1]

[1]MARUM - Centre for Marine Environmental Sciences, University of Bremen, Leobener Strasse, D-28359 Germany

[2]Geosciences Department, University of Bremen, Klagenfurter Strasse, D-28359 Bremen, Germany

E-mail: catalina@uni-bremen.de; dupont@uni-bremen.de

Abstract. A Cariaco Basin pollen record shows the development of tropical salt marshes during marine isotope stage 3. Rapid and abrupt expansions of salt marsh vegetation in tropical South America are associated with north Atlantic Heinrich Events stadials (HE-stadials). Intervals of salt marsh expansion have an internal structure, which consists of a recurrent alternation of species that starts with pollen increments of Chenopodiaceae, that are followed by increments of grasses, and subsequently by increments of Cyperaceae. This pattern suggests a successional process that is determined by the close relationship between sea-level and plant community dynamics. The salt tolerant Chenopodiaceae, indicate hypersaline intertidal environments, which were most likely promoted by extremely dry atmospheric conditions. Rapid sea-level rise characterizes the onset of HE-stadials, causing the continued recruitment of pioneer species, which are the only ones tolerating rapid rates of disturbance. Once sea-level rise decelerates, marsh plants are able to trap and stabilize sediments, favouring the establishment of more competitive species. These results add to the scarce knowledge on the dynamics of tropical salt marsh ecosystems, and provide independent paleoclimatic evidence on sea-level changes following Antarctic climate variability.

1. Introduction

The timing of sea level changes during marine isotope stage 3 (MIS 3; 60-25 ka) is a key issue in understanding the role of ice sheets in millennial-scale climate variability. The available reconstructions of sea level changes during this interval greatly rely on oxygen isotope records from deep-sea cores (since coral-based data are sparse and chronologies less precise), and consistently show four cycles of similar amplitude of sea level change in the order of 20-30 m [1, and references therein]. However, there is little agreement on the exact timing of these changes or on the relative roles of the Southern and Northern Hemisphere ice sheets in global sea level scenarios.

The ecological response of sensitive terrestrial ecosystems can provide independent information that complements the almost exclusively marine body of evidence of millennial sea level change. For this purpose, intertidal tropical ecosystems can be particularly useful, since they are very sensitive to environmental gradients in the sea-continent interface. In tidal salt marsh plant communities, species composition varies with elevation, usually in a banded pattern parallel to the shore. Its variation often reflects environmental gradients that result from the interaction between tidal regime, local topography, freshwater input, and biota. It has been proposed that the zonation is a spatial expression of successional changes over time and has potential to be reconstructed for the past by pollen analysis. If patterns of pollen deposition follow zonation and succession patterns, these can be reconstructed back in time by establishing a time-depth relationship with the fossil evidence, this then enables past sea level to be reconstructed. In this study, we present a palynological reconstruction from marine core MD03-2622, Cariaco Basin, in northernmost South America, that evidences the dynamics of intertidal plant communities during intervals associated with Heinrich Event stadials (HE-stadials), linking them to the well-constrained North Atlantic millennial- to sub-millennial climatic variability.

The Cariaco Basin is located on the northern shelf of Venezuela and is particularly sensitive to the seasonal shifts of the Intertropical Convergence Zone (ITCZ), which deeply influence the hydrology and oceanographic features of the basin. During MIS 3, the Cariaco Basin record displays a clear North Atlantic climatic variability, shifting from dry conditions during cold stadials to wet conditions and increased river runoff during warm interstadials. This hydrological pattern is reflected by variations in the input of terrestrial materials and has been explained by the latitudinal migration of the ITCZ [2, 3, 4]. The chronology used in this study was established by linking similar features of sediment reflectance profile of Cariaco site MD03-2622 with that of the nearby ODP Site 1002D, which has an extremely high-resolution age model for the past 60 ka based on more than 350 AMS ^{14}C data on planktic foraminifera [5, 6] (correlation coefficient between both chronologies is 0.74). Bigger uncertainties in the chronology arise before 40 ka, because of the differences between Hulu Cave chronology (used to tie Cariaco Basin chronology) and Greenland chronologies. Thus, the age model during the high resolution interval of this study is highly reliable.

2. Tropical salt marsh response to millennial climate and sea level changes

During glacial periods, when sea level was 80-120 m lower than today, a broad shallow shelf –the Unare platform- became exposed south of the Cariaco Basin. Periods of extremely dry atmospheric conditions might, therefore, have resulted in hypersaline coastal environments. These extreme conditions could have been tolerated only by a limited number of plant species. Chenopodiaceae, Poaceae and Cyperaceae belong to the most common representatives of salt tolerant plants in tropical and subtropical wetlands [7] (figure 1).

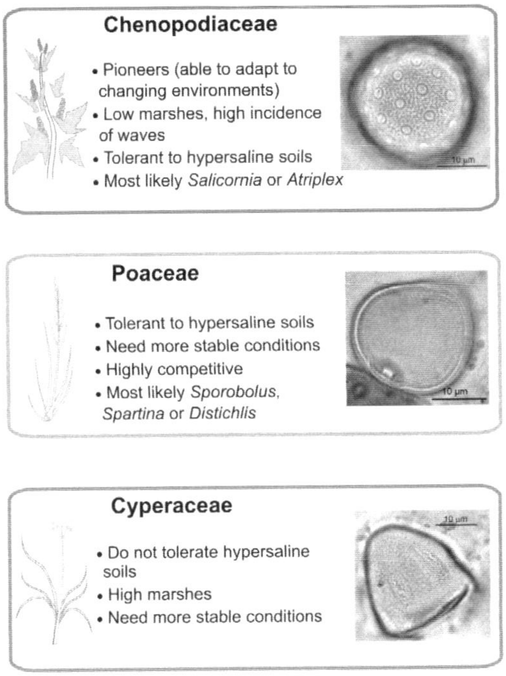

Figure 1. Ecological preferences of three salt marsh taxa.

2.1. The pollen record

Five high-amplitude vegetation shifts were recorded in the pollen record during MIS 3 (60-25 ka), indicating rapid oscillations of environmental conditions in northernmost South America. Recurrent salt-tolerant vegetation expansions were shown to correlate with HE stadials 3-6. Within single HE-stadials, a recurrent and directional succession of pollen taxa was observed in the following order: abrupt increases in saltbush (Chenopodiaceae) followed by a dominance of grasses (Poaceae), which in turn were replaced by sedges (Cyperaceae) (figure 2 and 3). Once interstadial conditions returned, mangroves and other arboreal species replace the herbaceous taxa.

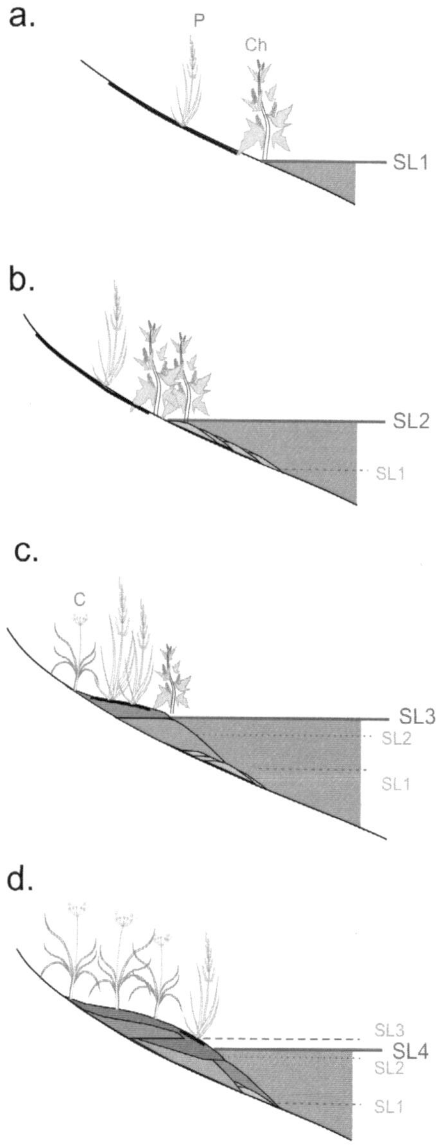

a. Establishment of salt marshes when arid conditions promote extensive hypersaline intertidal environments. Extended periods of strong evaporation and reduced rainfall occur during stadials HE stadials (Peterson and Haug, 2006; González et al., 2008).

b. Rapid sea level rise causes erosion; only pioneer species tolerate the rapid change. Intervals of maximum pollen representation of Chenopodiaceae are interpreted as periods of permanent tidal influence and sediment relocation, directly related to acelerated sea-level rise.

c. Sea level rise decelerates, and accretion of sediments and autochthonous organic material takes place; more competitive perennial grasses take advantage of favorable conditions. Areas covered with vegetation experienced higher increments of marsh height.

d. Sea level drops, sediment accumulation contrains the tidal influence to the seaward edge. The presence of Cyperaceae necessarily indicates less saline conditions. Since there is no evidence of increased freshwater input during HE-stadials, Cyperaceae pollen maxima might reflect an expansion of elevated marsh areas due to lowered sea levels.

Figure 2. Schematic representation of salt marsh community dynamics in a changing sea level environment according to the Cariaco Basin pollen record [11]. Thicker black lines indicate areas of soil hypersalinity. SL 1 to SL4 denote different sea levels reconstructed from the pollen record and correspond with phases indicated in figure 3.

Even if Chenopodiaceae, Poaceae and Cyperaceae can occur in diverse environments, several lines of evidence suggest that the succession indicated in the MD03-2622 pollen record during HE-stadials represents the expansion of salt-marshes on coastal environments, and thus can be used as sea-level indicator. The morphology of the pollen grains, the current distribution of the genera in the neighboring area of Cariaco Basin, and perhaps most convincing, the coincidence of the mangrove tree *Rhizophora* (only occurs intertidally) all suggest a gradational pattern indicative of intertidal habitats.

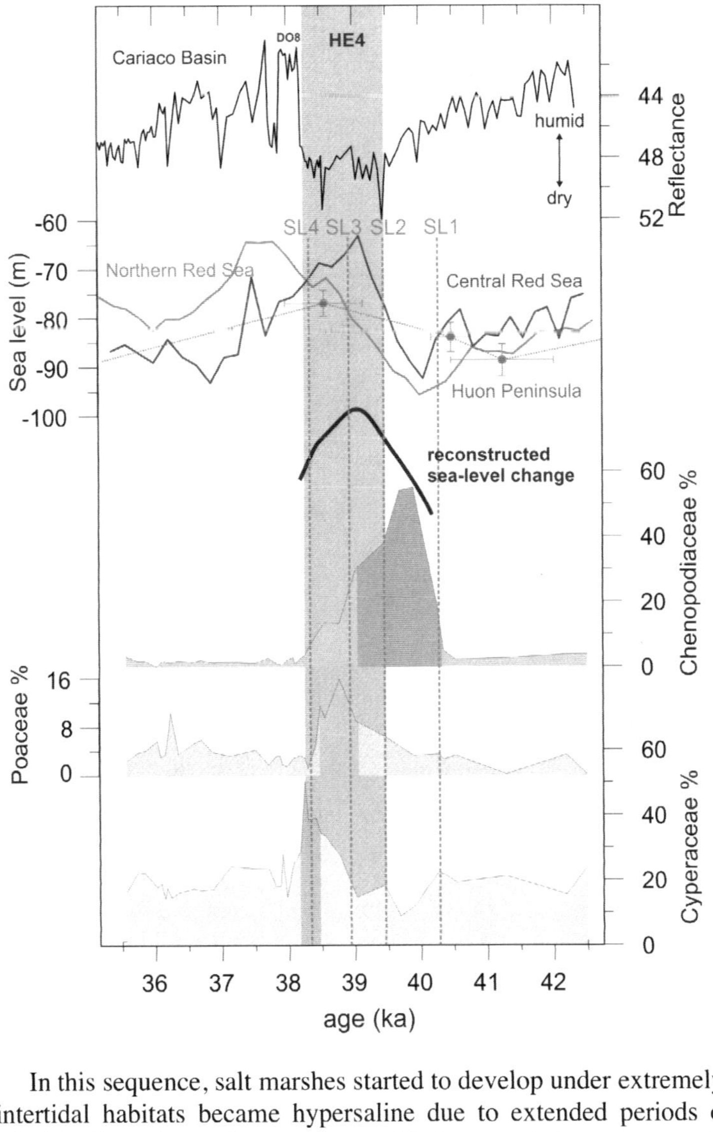

Figure 3. Comparison of the high-resolution palynological record from core MD03-2622 (Cariaco Basin) with sea level reconstructions from Red Sea marine sediment cores and Huon Peninsula (Papua New Guinea) fossil corals during HE-stadial 4 (González and Dupont, 2009). Top to bottom: Reflectance data from core MD03-2622 [18]. Sea level data; dark blue line - central Red Sea [1, 14], light blue line - northern Red Sea (Arz et al., 2007), and dotted pink line - Huon Peninsula [10]. Reconstructed direction of sea-level changes based on pollen reconstruction. Pollen relative abundance of Chenopodiaceae, Poaceae, and Cyperaceae [11] indicating the directional alternation of salt marsh species. Dotted gray lines SL1 to SL4 denote different sea levels reconstructed from the Cariaco Basin pollen record, which correspond to phases explained in figure 2.

In this sequence, salt marshes started to develop under extremely arid stadial conditions [3, 4] when intertidal habitats became hypersaline due to extended periods of strong evaporation and reduced rainfall. The salt marshes were most likely restricted to narrow intertidal areas because under strongly seasonal conditions they are usually fringed on the landward side by extensive bare salt pans (figure 2a, b; [7]). Early colonizing species of salt marshes, like the annual *Atriplex* and *Salicornia* (Chenopodiaceae), first colonize bare zones of lower and middle marsh areas, with a high incidence of waves and prolonged inundation regimes [8]. Thus, intervals of maximum pollen representation of Chenopodiaceae are interpreted as periods of direct tidal influence and sediment relocation. Frequent tidal flooding under accelerated sea level rise would result in flooding of the marsh surface, transforming it into a new seafloor, with the later landward accretion of new, low marsh sediments (figure 2b).

By comparing our high-resolution pollen data with sea level reconstructions from the Red Sea [1, 9 and references therein] and the independently dated fossil corals from the Huon Peninsula [10] for the period between 40.5-38 ka, we found that the phase dominated by Chenopodiaceae corresponds closely with an interval of accelerated sea level rise (figure 3). This confirms that only early successional plants, with high colonizing abilities (e.g., rapid growth, annuals or short-lived perennials) were capable of surviving the stressful high rates of change (figure 2a). Moreover, the erosion of low marsh sediments would wash out and transport the pollen produced in situ (figure 2b).

As soon as sea level rise decelerated (ca. 1 ka after the Chenopodiaceae peak), some vegetation was able to establish permanently. In low marsh areas, sediment accretion greatly depends on vegetation cover, which limits erosion, and enhances sediment and organic matter trapping. Thus, areas covered with vegetation experienced higher marsh heights. The build-up of middle and high marsh environments favoured the expansion of more competitive perennial grasses (Poaceae), thus replacing Chenopodiaceae pioneer species (figure 2c, d and 3). In contrast, the presence of Cyperaceae indicates less saline conditions, since sedges do not tolerate salinity excess. Thus, since there is no evidence of increased freshwater input during HE-stadials, Cyperaceae pollen maxima might reflect an expansion of elevated marsh areas (figure 2c, d).

Once interstadial conditions resumed and the average position of the ITCZ shifted northwards, the increased availability of freshwater might have alleviated salinity stress on soils, allowing a more complex plant community to develop on the shelf, and pushing the upper borders of the salt marsh seawards. Simultaneous increases in mangrove pollen [11] confirm that coastal environments became less saline and increasingly suitable for the establishment of forests during stadial-interstadial transitions. In addition to freshening, decelerated sea level rise (or sea level fall) would be required to allow the establishment of mangroves, since mangroves do not survive if sea level rise occurs too rapidly [12,13].

2.2. An alternative view of Chenopodiaceae?

Most of our interpretation of Cariaco pollen record in terms of sea-level change is based on the assumption that Chenopodiaceae peaks are indicating rapid sea-level rise episodes. However, due to the affinity of these plants for hypersaline soils, one could intuitively argue that, on the contrary, Chenopodiaceae indicate periods of decreasing sea-level and the corresponding expansion of suitable areas for colonization. In this respect, we should highlight first, that during dry periods, tropical salt marshes only occupy narrow intertidal areas in direct contact with seawater that are usually fringed on their landward side by extensive hypersaline vegetation-bare flats. As a result, with decreasing sea-levels, absolute size of bare areas would increase, but the integrated areas that are inundated by tides in fact decrease (for a bathymetry like that of Cariaco Basin where the perimeter of isobaths become smaller with increasing depth). Following the same reasoning, we estimated an increment of ~2200 km^2 of intertidal areas by the effect of rising sea-level from -90 to -60 m below sea-level (figure 4), that at a rate of 3 cm/yr [1] would imply an approximate gain of 2.3 km^2 of intertidal marsh surface area every year during ca. 1000 years.

Conversely, the counterargument that high percentages of Chenopodiaceae pollen are indicating lowering instead of rising sea levels might provide an alternative, valid explanation. By following the opposite logic, we would require an explanation to reconcile the succession of plant species with the expected sequence from sea-level changes. Under this alternative view, Chenopodiaceae would expand due to the rapid expansion of suitable areas. The lower the sea-level, the larger the suitable areas, and the more Chenopodiaceae plants that can grow. It is important to note that succession is not a symmetrical process, and what occurs on one direction does not necessarily occur in the opposite. When sea-level drops, secondary species like Poaceae and Cyperaceae can compete for resources with pioneers especially on higher marsh areas, where conditions are more stable (i.e. less erosion, accumulation of substrate, eventually less salinity). With a permanent drop in sea-level occurring along several hundreds of years, there is no reason for pioneers to dominate the pollen record and persist over secondary species, as it is the case in the Cariaco Basin record. Therefore, based on this reasoning, we conclude that the rapid change of sea-level indicated by the direction of the succession of species, supports rising and not lowering sea-levels, as one might possibly argue.

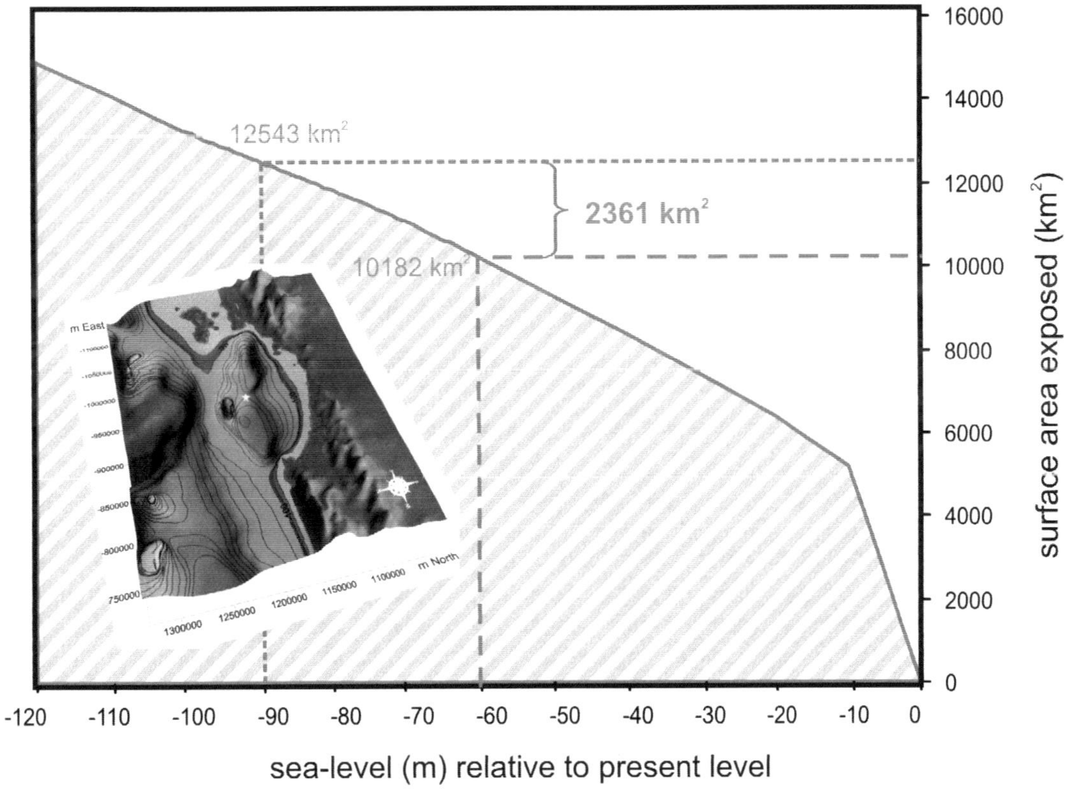

Figure 4. Estimated surface area (km^2) exposed in the Cariaco Basin at different sea-levels. The blue dotted lines indicate the increment of surface area by the effect of rising sea-levels from -90 to -60 m. On the lower corner, the bathymetric map of Cariaco Basin and adjacent areas. The red-shaded area denotes the increment of marsh surface area by 30 m of sea-level rise (from -90 to -60 m.b.s.l.).

3. Comparison and climatic implications

The Cariaco Basin pollen record also shows a similar relation between salt-marsh expansion and sea level rise during HEs 3, 5, 5a and 6, in spite of dating uncertainties and poorer resolution of the vegetation data [11]. In all five cycles, maximum values of Chenopodiaceae pollen coincide with the onset of HE stadials in the North Atlantic, and with warming phases in Antarctica. According to our palynological evidence, sea level started to rise before the ice sheet collapse that caused Heinrich layers in the North Atlantic, being in agreement with both Red Sea sea level reconstructions during the HE 4 (figure 3; [1, 9] and with fossil coral data from the Huon Peninsula [10]. However, a subsequent decelerated rise or fall of sea level is needed to reconcile with the expansion of Poaceae. In this case, our data supports the timing of central Red Sea reconstruction [1, 14, 15], the independently dated corals from the Huon Peninsula, and models, which suggest that melting in Antarctica might have accounted for a rise in sea level of about 20 m [15,16,17].

4. Conclusions

The palynological analysis in marine core MD03-2622 from northernmost South America, allowed the reconstruction of interstadial vegetation dynamics during MIS 3, suggesting rapid and repeated sea level changes. Five intervals of expanded salt marsh vegetation corresponded to the onset of HEs of the northern high latitudes and indicate periods of accelerated sea level rise in the tropical Atlantic. The close relationship between sea level rise and community dynamics is consistent with a resource-based mechanism of succession, where soil development and salinity gradients are the main factors determining plant succession patterns of coastal marshes. In this context, the Cariaco Basin palynological record is especially informative on the timing of sea level changes during MIS 3 and their connection with HEs, supporting the idea that sea level fluctuations followed Antarctica climate variability.

Acknowledgements

This work was supported by the Programme Alβan –the European Union Programme of High Level Scholarships for Latin America (Scholarship E04D047330CO) – and the Deutsche Akademische Austausch Dienst (DAAD)-Colfuturo Program. Data will be available in PANGAEA (www.pangaea.de).

References

[1] Siddall, M., Rohling E.J., Thompson, W.G. and Waelbroeck, C., 2008: Marine isotope stage 3 sea level fluctuations: Data synthesis and new outlook, *Reviews of Geophysics*, 46: RG4003, doi: 10.1029/2007RG00226.

[2] Peterson, L.C., Haug, G.H., Hughen, K.A. and Röhl, U., 2000: Rapid changes in the hydrologic cycle of the tropical Atlantic during the last glacial, *Science*, 290: 1947-1951.

[3] Peterson, L. and Haug, G., 2006: Variability in the mean latitude of the Atlantic Intertropical Convergence Zone as recorded by riverine input of sediments to the Cariaco Basin (Venezuela), *Palaeogeography Palaeoclimatology Palaeoecology*, 234: 97-113.

[4] González, C., Dupont, L.M., Behling, H. and Wefer, G., 2008: Neotropical vegetation response to rapid climate changes during the last glacial: palynological evidence from the Cariaco Basin, *Quaternary Research*, 69: 217-230.

[5] Hughen, K., Lehman, S., Southon, J., Overpeck, J., Marchal, O., Herring, C. and Turnbull, J., 2004: 14C activity and global carbon cycle changes over the past 50,000 years, *Science*, 303: 202-207.

[6] Hughen, K., Southon, J., Lehman, S., Bertrand, C. and Turnbull, J., 2006: Marine-derived 14C calibration and activity record for the past 50,000 years updated from the Cariaco Basin, *Quaternary Science Reviews*, 25: 3216–3227.

[7] Adam, P., 2002: Saltmarshes in a time of change, *Environmental Conservation*, 29: 39-61.

[8] Ranwell, D.S., 1972: *Ecology of salt marshes and sand dunes*, Chapman and Hall, 258 p.

[9] Arz, H.W., Lamy, F., Ganopolski, A., Nowaczyk, N. and Pätzold, J., 2007: Dominant Northern Hemisphere climate control over millennial-scale glacial sea-level variability, *Quaternary Science Reviews*, 26: 312–321.

[10] Thompson, W.G. and Goldstein, S.L., 2006: A radiometric calibration of the SPECMAP timescale, *Quaternary Science Reviews*, 25: 3207-3215.

[11] González, C. and Dupont, L.M., 2009: Tropical salt marsh succession as sea-level indicator during Heinrich events, *Quaternary Science Reviews*, 28: 939-946.

[12] Ellison, J.C., 1993: Mangrove retreat with rising sea-level, Bermuda, *Estuarine, Coastal and Shelf Science*, 37: 75-87.

[13] Woodroffe, C.D., 1999: Response of mangrove shorelines to sea-level change, *Tropics*, 8: 159-177.

[14] Siddall, M., Rohling, E.J., Almogi-Labin, A., Hemleben, Ch., Melschner, D., Schmelzer, I. and Smeed, D.A., 2003: Sea-level fluctuations during the last glacial cycle, *Nature*, 423: 853-858.

[15] Rohling, E.J. et al., 2008: New constraints on the timing of sea level fluctuations during early to middle marine isotope stage 3, *Paleoceanography*, 23: doi:10.1029/2008PA001617.

[16] Rohling, E.J., Marsh, R., Wells, N.C., Siddall, M. and Edwards, N., 2004: Similar melt-water contributions to glacial sea-level variability from Antarctic and northern ice sheets, *Nature*, 430: 1016-1021.

[17] Flückiger, J., Knutti, R. and Watts, J.W.C., 2006: Oceanic processes as potential trigger and amplifying mechanisms for Heinrich events, *Paleoceanography*, 21: PA2014, doi: 10.1029/2005PA001204.

[18] Laj, C., 2004: *Cruise Report: MD 132 –P.I.C.A.S.S.O. Images XI, Fortaleza-Baltimore-Brest, Mai Juin 2003*, Institut Polaire Francais Paul Emile Victor, 53p.

The First-Order Effect of Holocene Northern Peatlands on Global Carbon Cycle Dynamics

Yi Wang[1,2], Nigel T. Roulet[3], Steve Frolking[4], Lawrence A. Mysak[5], Xiaodong Liu[2], and Zhangdong Jin[2]

[1] Department of Geography and School of Global Studies, University of Sussex

[2] Institute of Earth Environment, Chinese Academy of Sciences

[3] Department of Geography and McGill School of Environment, McGill University

[4] Complex Systems Research Center, University of New Hampshire

[5] Department of Atmospheric and Oceanic Sciences, McGill University

E-mail: yi.wang@sussex.ac.uk

Abstract. Given the fact that the estimated present-day carbon storage of Northern Peatlands (NP) is about 300-500 petagram (PgC, 1 petagram = 10^{15} gram), and the NP has been subject to a slow but persistent growth over the Holocene epoch, it is desirable to include the NP in studies of Holocene carbon cycle dynamics. Here we use an Earth system Model of Intermediate Complexity to study the first-order effect of NP on global carbon cycle dynamics in the Holocene. We prescribe the reconstructed NP growth based on data obtained from numerous sites (located in Western Siberia, North America, and Finland) where peat accumulation records have been developed. Using an inverse method, we demonstrate that the long-term debates over potential source and/or sink of terrestrial ecosystem in the Holocene are clarified by using an inverse method, and our results suggest that the primary carbon source for the changes (sinks) of atmospheric and terrestrial carbon is the ocean, presumably, due to the deep ocean sedimentation pump (the so-called alkalinity pump). Our paper here complements ref. 1 by sensitivity tests using modified boundary conditions.

1. Introduction

Holocene carbon cycle dynamics over the past 8000 years before present (8 kyr BP) have been a puzzle to the paleoclimate community [2-9]. There have been long-term debates of different roles of terrestrial and oceanic carbon systems in the Holocene. Many efforts to better understand Holocene carbon cycle dynamics have been focused on the sea surface temperature changes [2, 5-7, 9-10], deep ocean adjustments [3, 5, 7, 11-12], natural sources of terrestrial carbon cycles [2, 5-7], and anthropogenic land-use and land-cover changes [13-15]. A common feature of these efforts is the neglect of the slow but persistent carbon accumulation of NP.

The end of the last deglaciation was accompanied by the initiation and growth of NP [16-25]. Carbon storage in present-day NP is about 300-500 PgC [17-18, 21-23, 26-28]. The main uncertainties to include the NP into players of Holocene carbon cycle dynamics are: 1) average basal dates of NP [21-22], 2) bulk carbon densities of NP [29-30], and 3) mean peat depths of NP [27, 31].

Here we make an effort to include the first-order effect of peatlands on global carbon cycle dynamics in the Holocene by prescribing an idealized development of NP, based on recent work of refs. 21-22, 24, and 27. Our inverse method was first developed in ref. 7 (see Section 3 and figure 1 for details). The fundamental bases of our approach are: 1) the global carbon budget is conserved under the timescale of interests; 2) there are two-way fluxes between land and atmosphere and between ocean and atmosphere; and 3) there is only one-way flux from land to ocean, which is more than two orders of magnitude smaller than the two-way fluxes. Our model simulation presented here complements ref. 1 by extending simulations with modified boundary conditions to test the sensitivity of former results.

2. Model and Experimental Designs

The ``Green" McGill Paleoclimate Model (hereafter MPM) is a six-component (atmosphere, ocean, sea ice, land surface, ice sheet and vegetation) Earth system Model of Intermediate Complexity. The atmosphere component is a sectorially averaged energy-moisture balance model. The ocean component is a 2.5-D zonally averaged model. The MPM has been developed to include biogeophysical feedbacks [7, 32] so that the energy and moisture fluxes and interactions between land surface and atmosphere are considered interactively. However, the biogeochemical feedbacks have not been considered interactively in the model because of the absence of marine biogeochemistry (The carbon cycle is not closed in the model.). The ``Green" MPM has been used in studies of Holocene climate changes and carbon cycle dynamics [7, 33-34] with a retreating Laurentide Ice Sheet. The MPM also has been used to study the last glacial inception and glacial abrupt climate changes [35-37],

Following the approach in refs. 33-34, we prescribed a retreating Laurentide Ice Sheet from 8 to 6 kyr BP in our simulations. The terrestrial carbon cycle module is based on VECODE [5]. VECODE simulates the carbon stored in live vegetation and soil when forced by the climatic conditions (e.g., temperature and precipitation) in the model. The atmospheric CO_2 concentration used in VECODE will have only a fertilizing effect on the net primary productivity. In addition to the atmospheric CO_2, annual mean temperature and precipitation control the value of annual net primary productivity. The pre-industrial, mid-Holocene (6 kyr BP), and 8 kyr BP terrestrial carbon cycles in the ``Green" MPM have been validated in ref. 6, and are in good agreement with other modeling studies [5-6] and observations [38-39].

The atmosphere, land surface, and sea ice components of the ``Green" MPM, together with VECODE, are first spun up for 60 years under the forcings of (1) monthly mean insolation at the top of the atmosphere at 8 kyr BP [40], (2) the zonally averaged monthly sea surface temperature climatology, and (3) the reconstructed atmospheric CO_2 concentration from the Taylor Dome ice core (260 ppmv). This spin-up represents the forcing conditions and a quasi-equilibrium state at 8 kyr BP [7, 33-34, 41]. The ocean component is next coupled with the aforementioned four components using flux adjustments. The final equilibrium state of 8 kyr BP is reached after an additional 5000 years of integration. After the 5060-year spin-up, the ``Green" MPM is subject to an 8000-year transient simulation under the forcings of varying insolation, the reconstructed atmospheric CO_2 concentration from the Taylor Dome ice core, and a prescribed retreating of Laurentide Ice Sheet. The development of Holocene NP is prescribed as a separate and inert carbon reservoir.

3. An Inverse Method and Prescribed NP Developments

Because we do not have the marine biogeochemistry module in the ``Green" MPM, we have applied the inverse method as developed in ref. 7 (figure 1). Figure 1a illustrates the schematic diagram of pre-industrial carbon cycle dynamics. On the timescale of interest, there are three major players in the global carbon cycle, i.e. without considering the carbon exchanges between rocks and the atmosphere and the anthropogenic land-use and land-cover changes. Furthermore, if we exclude the one-way discharge from land to ocean as in studies of refs. 5-6, we can demonstrate that when both the land and atmosphere carbon storages increase (positive signs in figure 1b), the only source for both of those increases must come from the ocean (negative sign in figure 1b). This is mainly because the ocean could not directly supply carbon to land without the help of the atmosphere. Under other circumstances, we cannot uniquely determine the source of the atmospheric carbon increase without additional assumptions [7, 41]

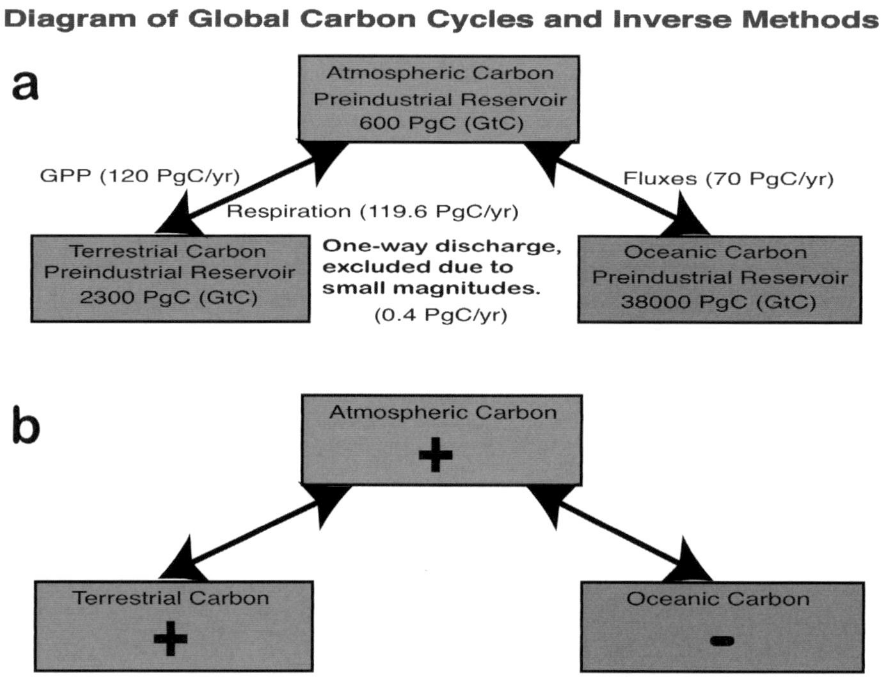

Figure 1. The schematic diagram of global carbon cycle dynamics with estimated pre-industrial storages and fluxes from the recent Intergovernmental Panel on Climate Change report (a), and an example of our inverse method showing the relative changes of carbon storage in three major reservoirs (b). The positive (negative) sign corresponds to carbon sink (source) in the global system.

NP have accumulated carbon through the Holocene [21-23, 42] with estimated present-day storages of 300-500 PgC [e.g., 17-18, 26]. Recent studies have advanced our understanding of Holocene NP dynamics in Western Siberia [16], North America [19, 24], and Finland [27]. Figure 2a shows the relative basal date frequencies (percentage) and accumulated percentages for the above-mentioned three regions as derived from those studies. To generate a weighted-averaged total percentage of accumulated basal-date (radiocarbon dates of the deepest/oldest peat in a peat core) frequency, we have assumed a relative area weighting of 10:5:1 for Western Siberia, North America, and Finland [42], respectively. To simplify our prescribed NP development (figure 2b) during the Holocene, we have assumed that (1) NP expansion is proportional to observed basal date frequency, and the total area is thus calculated as a linear function of basal dates (figure 2a, total frequency); and (2) the uptake rate per unit area is and has been constant, and thus the total carbon uptake scales linearly with peatland area [42]. Although in reality, NP growth is nonlinear and many factors have controlled its development through the Holocene, the slow but persistent carbon uptake from NP is a reasonable simple assumption. In particular, we prescribed three developing curves (figure 2b), which correspond to three estimated present-day NP carbon storages of 500, 400, and 300 PgC.

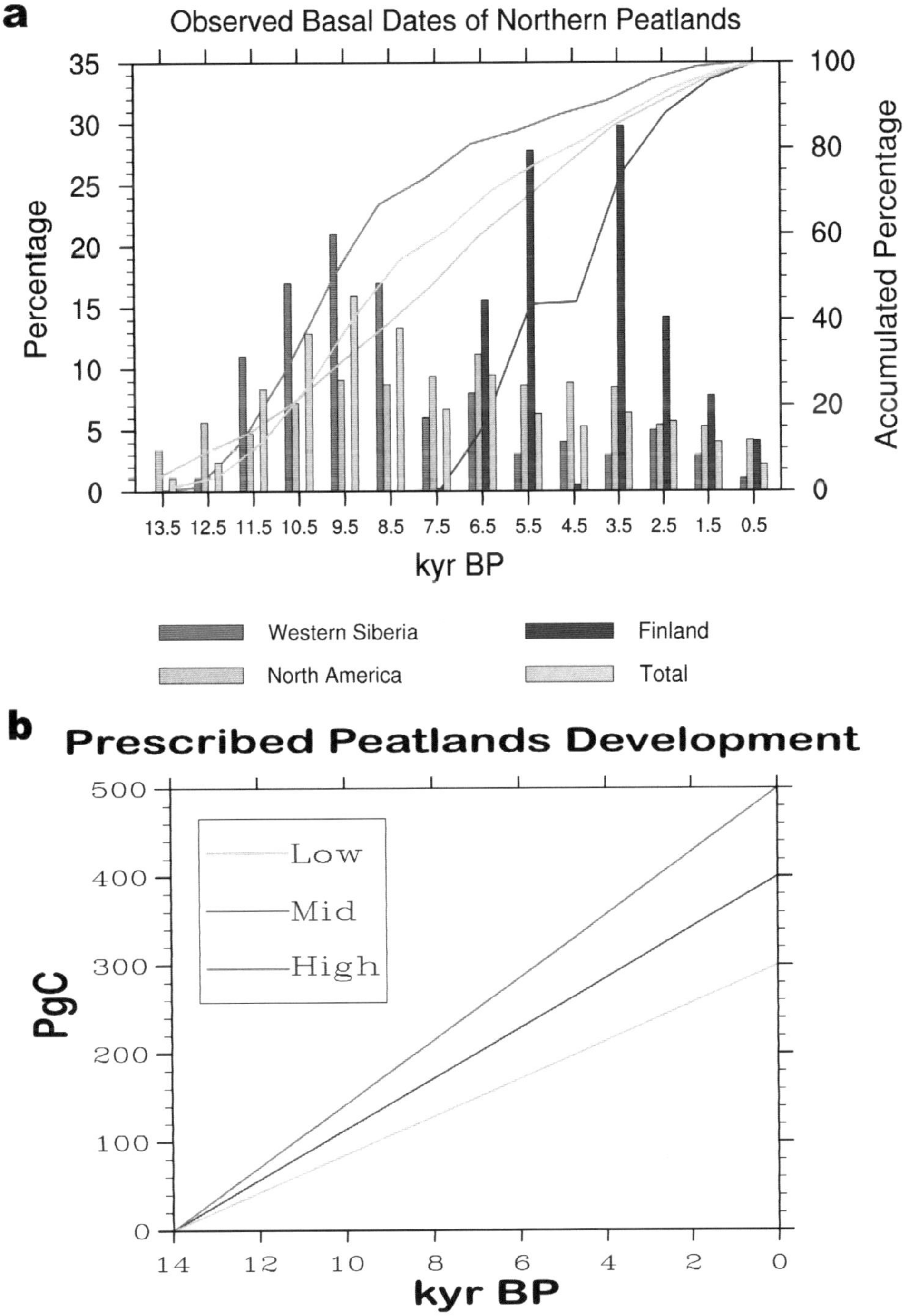

Figure 2. The observed basal date frequency of the NP for Western Siberia (red), North America (light blue), Finland (blue), and total (green) (a), and three idealized, prescribed NP development curves in our modelling study (b). Notice that the solid curves in (a) correspond to accumulative percentages of the observed basal date frequency with the same colour coding (right Y-axis).

4. Results

Here we only focus on the effect of the Holocene NP development on global carbon cycle dynamics (see detailed results in ref. 1). Ref. 7 indicated that from 8 to 6 kyr BP, both terrestrial and atmospheric carbon storages were increasing, and the only source they could determine for those increases was the ocean. Because of the late-Holocene desertification of the Sahara and boreal forest retreats, they could not uniquely determine the source of the atmospheric carbon increase from 6 to 0 kyr BP. Additional assumptions were made to partition the source of atmospheric carbon into the land and the ocean in ref. 7. However, if the slow but persistent development of NP is considered, the small release of carbon from the Sahara and boreal forest diminishes. Figure 3 shows estimates of the relative carbon storage changes in land (VECODE plus prescribed NP trajectories, green curves), atmosphere (ice core data, blue curves), and ocean (red curves, assuming global carbon conservation during the Holocene). Based on the inverse method, it is clear that, without considering anthropogenic land-use and land-cover changes, the effect of NP has been to sequester a large amount of carbon from both the atmosphere and the ocean. This sequestration makes it clear that over the past 8000 years, both land and atmosphere carbon storages have increased, and the sole source of these increases is the ocean. Because the global sea surface temperature only varies about 0.2°C in our model simulation (figure not shown here, but in ref. 1) and observations (see, e.g., ref. 10), we further suggest that the deep ocean sedimentation pump (through calcite carbonate compensation) is the main source for land and atmospheric carbon increases if we do not include the coral reef buildup theory [43] and the anthropogenic hypothesis [13].

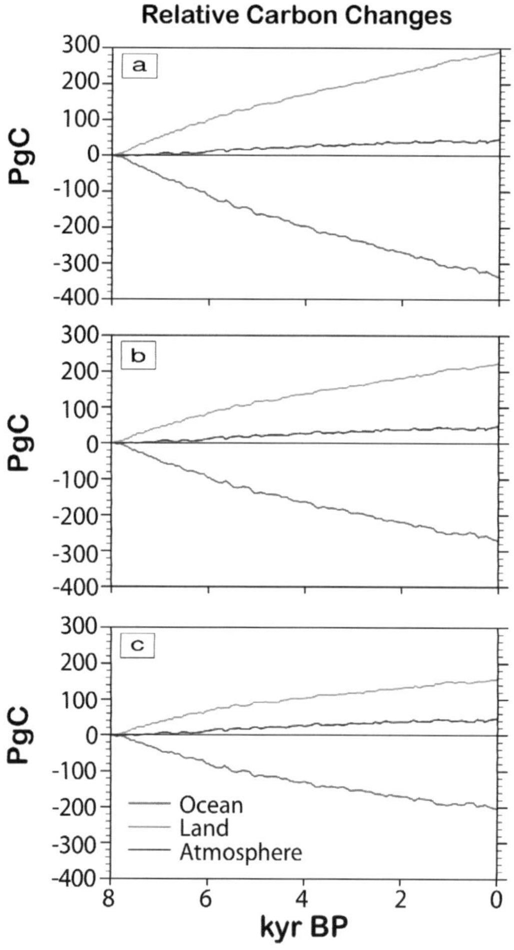

Figure 3. The relative carbon storage changes of land (VECODE and prescribed NP), atmosphere (ice core retrievals), and ocean (inferred assuming global carbon conservation). (a) corresponds to the red curve in figure 2b, (b) corresponds to the blue curve in figure 2b, and (c) corresponds to the green curve in figure 2b.

5. Summary and Future Work

Previous studies [2-8] have not been able to unambiguously identify the source of the atmospheric carbon changes in the Holocene (up to 8 kyr BP). However, based on three prescribed linear growths of NP (figure 2b), we have found, without additional assumptions, that the ocean is the sole source of the 20 ppmv CO_2 increase in the atmosphere over the past 8000 years. Our study provides the first-order effect of NP in Holocene carbon cycle dynamics. However, there are uncertainties in our current study, which provides the guidance for future studies of peatlands and Holocene carbon cycle dynamics.

The uncertainties related to our model simulations are mainly caused by the terrestrial vegetation responses to climatic forcing in both boreal forest and the Sahara/Sahel regions. For example, VECODE as an empirically-based land biogeochemistry module has large uncertainties in its simulation of the abrupt ecosystem collapse in mid-Holocene North Africa [44-46]. However, these uncertainties are not critical to the interpretation of our results, as the overall magnitude of NP carbon uptake is large. In addition, the radiative forcing of atmospheric CO_2 concentration is considered, but still contains large uncertainties in the model, which may change our results for the overall sea surface temperature variation in the Holocene.

Our prescribed NP development is highly idealized and is based on the present-day estimated total carbon storage of NP, and previous studies of NP initiation dates. However, large regions of NP do not have well-constrained basal dates. Even with well-constrained basal dates, carbon accumulation rates have varied considerably in many peatlands during the Holocene in response to varying climate conditions [47-48]. While the realistic NP development is unavailable and uncertain, the most significant effect of its growth is the uptake of carbon, and this effect changes the long-term debates regarding the potential carbon source from terrestrial ecosystem.

Our sensitive tests here show that under modified boundary conditions, the results of ref. 1 are robust. Therefore, NP development played an important role in Holocene carbon cycle dynamics. The centerpiece of the Holocene anthropogenic hypothesis is to claim that humans took control of atmospheric CO_2 and CH_4 trends thousands of years ago because of perturbations from land-use and land-cover changes [13-15]. However, without constrained magnitudes of those changes, it is still difficult to add this hypothesis into our model simulations. Nevertheless, our fundamental goal is to introduce interactively-coupled marine carbon dynamics and a dynamic peatland module so that the biogeochemical feedbacks (interactions) can be considered in future simulations of global carbon cycle dynamics in the Holocene.

Acknowledgments

During the course of our research, we have benefited from encouragement and discussion with Drs. C. Long, T. Kiefer, S. Rupper, W. Ruddiman, and Z. Yu. Our research has been supported by the Canadian Foundation for Climate and Atmospheric Sciences (CFCAS), the Natural Sciences and Engineering Research Council of Canada (NSERC), and the U.S. National Science Foundation (NSF ATM-0628399). X. Liu is supported by the NSFC National Excellent Young Scientists Fund (40825008). Z. Jin is supported by the National Basic Research Program Grant (2010CB833404). We thank two reviewers for their constructive comments.

References

[1] Wang Y, Roulet N T, Frolking S and Mysak L A 2009 *Climate of the Past* **5** 683-693
[2] Indermühle *et al.* 1999 *Nature* **398** 121-126
[3] Broecker W S, Lynch-Stieglitz J, Clark E, Hajdas I and Bonani G 2001 *Geochemistry Geophysics Geosystems* **2** doi: 10.1029/2001GC000177
[4] Monnin *et al.* 2001 *Science* **291** 112-114
[5] Brovkin V, Bendtsen J, Claussen M, Ganopolski A, Kubatzki C, Petoukhov V and Andreev A 2002 *Global Biogeochemical Cycles* **16** doi: 10.1029/2001GB001662
[6] Joos F, Gerber S, Prentice I C, Otto-Bliesner B L and Valdes P J 2004 *Global Biogeochemical*

Cycles **18** doi: 10.1029/2003GB002156
[7] Wang Y, Mysak L A and Roulet N T 2005 *Global Biogeochemical Cycles* **16** doi: 10.1029/2005GB002484
[8] Luthi D *et al.* 2008 *Nature* **453** 379-382
[9] Elsig J, Schmitt J, Leuenberger D, Schneider R, Eyer M, Leuenberger M, Joos F, Fischer H and Stocker T F 2009 *Nature* **461** 507-510
[10] Brovkin V, Kim J, Hofmann M and Schneider R 2008 *Global Biogeochemical Cycles* **22** doi: 10.1029/2006GB002885
[11] Broecker W S, Clark E, McCorkle D C, Peng T H, Hajdas I and Bonani G 1999 *Paleoceanography* **14** 744-752
[12] Broecker W S and Clark E 2003 *Global Biogeochemical Cycles* **17** doi: 10.1029/2002GB001985
[13] Ruddiman W F 2003 *Climatic Change* **61** 261-293
[14] Ruddiman W F 2006 *EOS* **87** 352
[15] Ruddiman W F 2008 *Quaternary Science Reviews* **27** 445-448
[16] Gorham E 1957 *Quaterly Review of Biology* **32** 154-166
[17] Gorham E 1991 *Ecological Application* **1** 182-195
[18] Gorham E 1995 *Biotic Feedbacks in the Global Climatic System: Will the Warming Speed the Warming?* Edited by Woodwell G and MacKenzie F (New York Oxford Univ. Press) pp 169-187
[19] Halsey L A, Vitt D H and Bauer I K 1998 *Climatic Change* **40** 315-342
[20] Gajewski K, Viau A, Sawada M, Atkinson D and Wilson S 2001 *Global Biogeochemical Cycles* **15** 297-310
[21] Smith L C, MacDonald G M, Velichko A A, Beilman D W, Borisova O, Frey K, Kremenetski K V and Sheng Y 2004 *Science* **303** 353-356
[22] MacDonald G M, Beilman D W, Kremenetski K V, Sheng Y, Smith L C and Velichko A A 2006 *Science* **314** 285-288
[23] Roulet N T, Lafleurs P M, Richard P J H, Moore T R, Humphreys E R and Bubier J 2007 *Global Change Biology* **13** 397-411
[24] Gorham E, Lehman C, Dyke A, Janssens J and Dyke L 2007 *Quaternary Science Reviews* **26** 300-311
[25] Yu Z, Beilman D W and Jones M C 2009 *Sensitivity of northern peatland carbon dynamics carbon dynamics to Holocene climate change* (American Geophysical Union Washington D.C., USA: AGU Geophysical Monographs) chap Carbon Cycling in Northern Peatlands doi:10.029/2008GM000822
[26] Roulet N T 2000 *Wetlands* **20** 605—615
[27] Turunen J, Tomppo E, Tolonen K and Reinikainen A 2002 *Holocene* **12** 69-80
[28] Limpens J, Berendse F, Blodau C, Canadell J G, Freeman C, Holden J, Roulet N T, Rydin H and Schaepman-Strub G 2008 *Biogeosciences* **5** 1475--1491
[29] Barber K E, Chambers F M and Maddy D 2003 *Quaternary Science Reviews* **22** 521-539
[30] Yu Z 2006 *Ecosystems* **9** 1278—1288
[31] Gorham E, Janssens J A and Glaser P H 2003 *Canadian Journal of Botany* **81** 429-438
[32] Wang Y, Mysak L A, Wang Z and Brovkin V 2005 *Climate Dynamics* **24** 469-480
[33] Wang Y 2005 *Simulation of the climate, ocean, vegetation and terrestrial carbon cycle in the Holocene* Ph.D. thesis McGill University Department of Atmospheric and Oceanic Sciences, McGill University, 805 Sherbrooke Street West, Montreal, Quebec, Canada H3A 2K6 available as Centre for Climate and Global Change Research C^2GCR Report No. 2005-3
[34] Wang Y, Mysak L A, Wang Z and Brovkin V 2005 *Climate Dynamics* **24** 481-496
[35] Wang Z, Cochelin A S, Mysak L A and Wang Y 2005 *Geophysical Research Letters* **32** doi: 10.1029/2005GL
[36] Mysak L A 2008 *Atmosphere-Ocean* **46** 317-341

[37] Wang Z and Mysal L A 2006 *Paleoceanography* **21** doi: 10.1029/2005PA001238
[38] Schimel D S, Braswell B H, Holland E A, McKeown R, Ojima D S, Painter T H, Parton W J and Townsend A R 1994 *Global Biogeochemical Cycles* **8** 279-293
[39] Falloon P, Smith P, Colemand K and Marshall S 1998 *Soil Biology and Biochemistry* **30** 1207-1211
[40] Berger A L 1978 *J. Atmos. Sci.* **35** 2362-2367
[41] Wang Y and Mysak L A 2005 *Geophysical Research Letters* **32** doi: 10.1029/2005GL023344
[42] Frolking S and Roulet N T 2007 *Global Change Biology* **13** 1079-1088
[43] Ridgwell A J, Watson A J, Maslin M A and Kaplan J O 2003 *Paleoceanography* **18** doi: 10.1029/2003PA000893
[44] Liu Z, Wang Y *et al.* 2007 *Quaternary Science Reviews* **26** 1818-1837
[45] Wang Y, Notaro M, Liu Z, Galimore R, Levis S and Kutzbach J~E 2008 *Climate of the Past* **4** 59-67
[46] Notaro M, Wang Y, Liu Z, Gallimore R and Levis S 2008 *Global Change Biology* **14** 347--368
[47] Yu Z, Campbell I D, Campbell C, Vitt D H, Bond G C and Apps M J 2003 *The Holocene* **13** 801-808
[48] Belyea L R and Malmer N 2004 *Global Change Biology* **10** 1043--105

Assessing antiquity and turnover of terrestrial ecosystems in eastern North America using fossil pollen data: A preliminary study

Yao Liu[1*], Stephen T. Jackson[1], Simon Brewer[2], John W. Williams[3]

1 Department of Botany and Program in Ecology, University of Wyoming, USA

2 Department of Botany, University of Wyoming, USA

3 Department of Geography and Center for Climatic Research, University of Wisconsin, USA

*yliu11@uwyo.edu

Abstract. We explored formal approaches to identifying and interpreting the antiquity and turnover of terrestrial ecosystems in eastern North America using pollen records. Preliminary results of cluster analyses, receiver-operating characteristic (ROC) analyses, and likelihood estimation of ecosystem analog in a simple Bayesian model allow assessment of modern ecosystem antiquities and past ecosystem turnovers. Approaches discussed in this study thus provide a vehicle for further studies.

1. Introduction

The emergence and disappearance of terrestrial ecosystems are driven by the turnover of their dominant species or functional types. Defined by their dominant plant species or types, ecosystems vary widely in their antiquity, with some formed thousands of years ago and others arising in the last few centuries [1, 2]. Ecosystem properties control many important land-surface processes, with consequences for hydrology [3], biogeochemistry [4, 5], carbon sequestration [6], and local to regional climate dynamics [7, 8]. Understanding of ecosystem antiquity is therefore important in assessing biosphere-atmosphere feedbacks and climate-change impacts [9, 10, 11]. Furthermore, antiquity and turnover patterns of specific ecosystems may provide information on their environmental thresholds and resilience in a changing climate [2, 12]. Fossil pollen records, which are widespread in eastern North America [1], provide a basis for assessing antiquity and turnover of modern and past ecosystems (Figure 1).

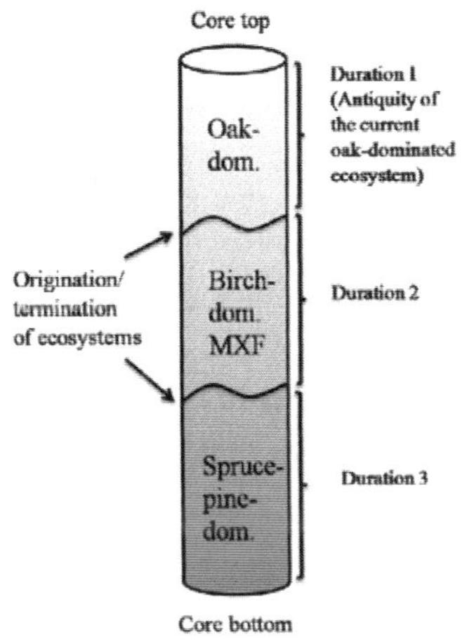

Figure 1. Conceptual figure showing the antiquity, turnover, and durations of ecosystems based on a fossil pollen sequence from a sediment core. The pollen abundances vary continuously throughout the core. Ecosystems (spruce-pine dominated, birch-dominated mixed forest and oak-dominated ecosystems) are categorized by their dominant pollen taxa.

Identifying and interpreting ecosystem antiquity and turnover from pollen records, however, can be challenging for three reasons. First, studies using modern analog techniques to investigate long-term vegetation change have usually imposed a single dissimilarity threshold (e.g. [13, 14, 15]). In reality, optimal dissimilarity threshold values vary across ecological scales and among ecosystems [16, 17]. Second, the categorical representation of the vegetation does not account for the temporal and spatial heterogeneity within an ecosystem category [1], and therefore may not represent the gradualness or abruptness of transitions in temporal vegetation composition. Third, constrained ordination techniques for pollen zonation in previous studies, such as CONISS [18], assess all vegetation transitions in a generalized way. Different ecosystems, however, may turn over in different ways, involving different mechanisms, rates, and manifestations in pollen records. Interpretation of ecosystem turnover should be assessed on the basis of individual ecosystems rather than all vegetation types as a whole. Recently developed analytical methods provide tools to address these challenges.

We identified and mapped the antiquity of modern ecosystems at individual fossil pollen sites (Figure 1) in eastern North America. Our objectives were to explore the utility of receiver-operating characteristic (ROC) analysis in a Bayesian framework in identifying ecosystem antiquity and turnover, and to assess spatial and temporal patterns of origination of modern ecosystems. In this paper, we describe our approaches and present results of preliminary analyses.

2. Data and analyses

2.1. Data

The modern calibration dataset used in this study were obtained from the North American surface sample dataset (http://www.geography.wisc.edu/faculty/williams/lab/Downloads.html, [19, 20]). Fossil pollen records were drawn from the NEOTOMA database (http://www.neotomadb.org). We used fossil pollen records from the NEOTOMA database that met the following criteria: 1) east of 105° W, 2) default age-depth model as chosen by the NEOTOMA team, 3) time-span greater than 1000 years, and 4) minimum sample density greater than one sample per 2000 years. The latter filter is admittedly coarse, but was necessary to ensure a broad geographic coverage of sites. A total of 479 fossil pollen records from sites in eastern North America were used in this study. All analyses of modern and fossil pollen data used a list of 25 pollen taxa common in eastern North America [15]: *Abies*, *Acer*, *Alnus*, *Betula*, *Carya*, *Corylus*, Cupressaceae/Taxaceae, Cyperaceae, *Fagus, Fraxinus*, *Juglans*, *Larix*, *Liquidambar*, *Ostrya/Carpinus*, *Picea*, *Pinus*, *Quercus*, *Platanus*, Poaceae, *Populus*, prairie forbs (Chenopodiaceae/Amaranthaceae, *Artemisia*, and other Asteraceae excluding *Ambrosia*), *Salix*, *Tilia*, *Tsuga*, and *Ulmus*.

IOP Conf. Series: Earth and Environmental Science **9** (2010) 012005 doi:10.1088/1755-1315/9/1/012005

2.2. Ecosystem classification: The rationale of applying cluster analyses

Terrestrial ecosystems can be defined by relative abundances of dominant taxa [21, 22, 23, 24]. In this work, we provisionally defined modern ecosystems by their dominant taxa, as represented by pollen percentages [25].

Before conducting a receiver operating characteristic (ROC) analysis, the modern calibration pollen dataset need to be classified into categories. There are at least two potential approaches to classify vegetation at a regional scale: 1) applying existing classification such as IGBP land-cover classification [26], or 2) conducting a cluster analysis. Here we compared the performance of both classifications in ROC analyses.

For the IGBP land-cover classification approach, we used the IGBP21 classification of surface pollen sites compiled in the North American surface sample dataset [19, 20]. For the cluster analysis approach, first we obtained the square root of all pollen percentages to down-weight overrepresented taxa and up-weight underrepresented taxa [27]. Then we conducted a set of k-means [28] cluster analyses using a range of values (from 1 to 30) of the total number of cluster centers. From these, we identified 15 modern "pollen-ecosystems", a number that minimized the ratio of within-cluster variation to across-cluster variation [29]. These palynologically defined ecosystems have distinctive pollen signatures. Four such ecosystems are shown in Figure 2. The grassland ecosystem (Figure 2a), the spruce-dominated ecosystem (Figure 2b), the pine-dominated mixed forest (Figure 2c), and the birch-dominated mixed forest (Figure 2d) correspond with, respectively, the eastern portion of the grassland [30], the spruce-dominated boreal forest, the western portion of mixed forest, and the eastern portion of mixed forest [31].

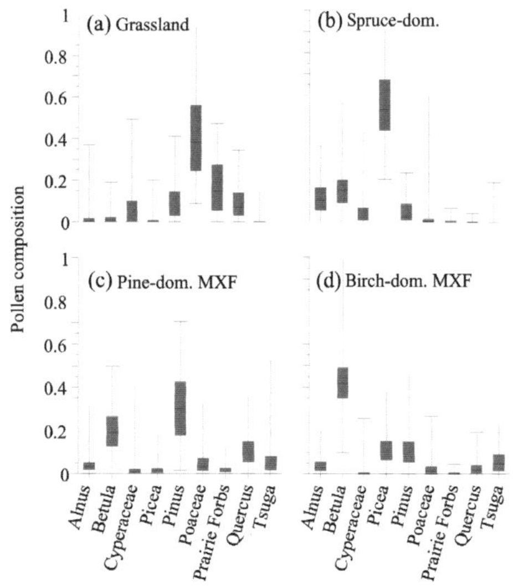

Figure 2. Examples of identified ecosystem types, shown by box-plots of percentages of nine representative pollen taxa that are significant in differentiating these and other "pollen ecosystems". a) Grassland ecosystem, b) Spruce-dominated ecosystem, c) Pine-dominated mixed forest ecosystem, d) Birch-dominated mixed forest ecosystem.

We then conducted two sets of receiver operating characteristic (ROC) analyses on both classified modern pollen datasets to obtain ecosystem-specific dissimilarity optimal thresholds for distinguishing same-ecosystem pollen samples from different-ecosystem pollen samples [17]. The square-chord distance (SCD) was used to quantify dissimilarity between two pollen assemblages.

The area under the curve (AUC) is a measure of the overall discriminative ability of a dissimilarity threshold. We compared the AUCs from two sets of ROC results at discriminating same-ecosystem pollen samples and different-ecosystem pollen samples. We found that AUCs were more than 20% higher in the clustering approach than the IGBP classification approach. This probably resulted from pollen-assemblage sensitivity to subregional vegetation patterns unrepresented in the coarser-scale IGBP classification [32].

We therefore chose to use the classification given by the clustering approach over the IGBP classification approach (used in [17]) for further analyses. In the rest of the paper, we will focus on presenting results of antiquity and turnover behaviors for the four ecosystems showed in Figure 2.

2.3. Constructing Bayesian likelihood of analog for fossil records

For a fossil pollen assemblage, the likelihood ratio of having a modern analog is defined as the ratio of being the same ecosystem versus being a different ecosystem given a SCD. For each fossil pollen sample, we computed this likelihood ratio of having an analog from each classified modern ecosystem.

First, we calculated the SCD score for each sample in a fossil record to the most similar modern pollen sample (palynologically, the sample with the smallest metric distances in the modern pollen dataset) from each modern ecosystem group was calculated. If this SCD was smaller than the ecosystem-specific optimal threshold calculated in Section 2.2, the likelihood ratio of the modern sample being an analog to the fossil sample analog was computed.

To estimate the likelihood of analog and the uncertainty associated with it, we fitted a power function curve that describes the relationship between the likelihood ratio and the SCD (Equation 1) in a Bayesian framework. As reported by Gavin *et al.* [17], we observed a highly non-linear relationship between the likelihood ratio (probability of analog versus no-analog) and the SCD in the classified modern calibration dataset (Figure 3).

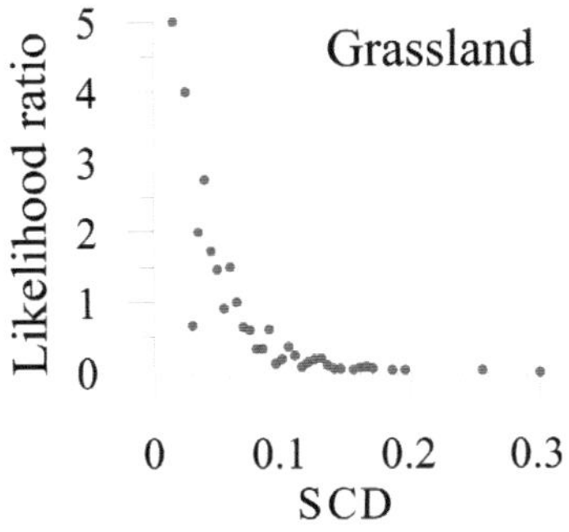

Figure 3. Scatter plot of likelihood ratio of an analog to grassland against SCD score to the nearest neighbor classified as modern grassland.

In Equation 1, $LR_{i,j}$ is the likelihood ratio of the probability that pollen assemblage *i* has an analog in ecosystem *j* verses the probability that pollen assemblage *i* has no analog in ecosystem *j*. $SCD_{i,j}$ is the SCD score of sample *i* to its nearest neighbor (sample with the smallest SCD) in ecosystem *j*. a_j and b_j are ecosystem-specific parameters for ecosystem *j*.

$$LR_{i,j} = a_j \times SCD_{i,j}^{b_j} \qquad \text{Equation 1}$$

We calculated all the $LR_{i,j}$ and $SCD_{i,j}$ using the modern calibration pollen dataset. Then we estimated the posterior distribution of ecosystem-specific parameters with a simple Bayesian model. The model fitting was based on the least sum of squares on log scale. The posteriors of fitted parameters were then used in the same model to compute likelihood ratios (probability of analog versus no-analog) with Bayesian confidence intervals for fossil pollen samples given their SCDs to modern samples.

2.4. Identifying antiquity and turnover

At each individual site, we estimated the antiquity of a modern ecosystem as the time elapsed since the pollen-analog of this modern ecosystem appeared and persisted in the record. A transition between ecosystem types was identified when the confidence intervals of likelihood-of-analog no longer overlapped. We mapped antiquities for the grassland ecosystem, the spruce-dominated ecosystem, the pine-dominated mixed forest ecosystem, and the birch-dominated mixed forest ecosystem.

IOP Conf. Series: Earth and Environmental Science **9** (2010) 012005 doi:10.1088/1755-1315/9/1/012005

3. Ecosystem antiquity and turnover in eastern North America

3.1. Antiquity and turnover at individual sites

Antiquity and turnover were assessed for each fossil pollen site. Results of analyses on the pollen sequence from Tannersville Bog [33] provide an illustration (Figure 4). For the antiquity of the modern ecosystem, the current oak dominated ecosystem at Tannersville Bog originated 10.5 ka BP. At this time, fossil samples start to have oak-dominated ecosystem modern analogs, which persist to the present.

Besides the uncertainty associated with the age model, an ecosystem transition is often a bracketed time period rather than a time point due to 1) sample density, 2) presence of no-analog ecosystems, and 3) persistence of antecedent ecosystems in a mosaic with adventive ecosystems, shown by overlapping confidence intervals of likelihood estimates for different ecosystems. A transition time period can range from less than a hundred years up to more than a thousand years. At Tannersville Bog, two major ecosystem transitions appeared to have happened in the pollen record (Figure 4). The transition from a spruce-pine dominated ecosystem to a birch-dominated mixed forest ecosystem occurred between 15.5 ka BP and 14 ka BP. Samples during this transition appeared to have no analogs in modern ecosystems. Later in the record, the transition from a birch-dominated mixed forest ecosystem to an oak dominated ecosystem, which spanned five adjacent pollen samples with overlapping confidence intervals of likelihood, happened between 11 ka BP and 10 ka BP.

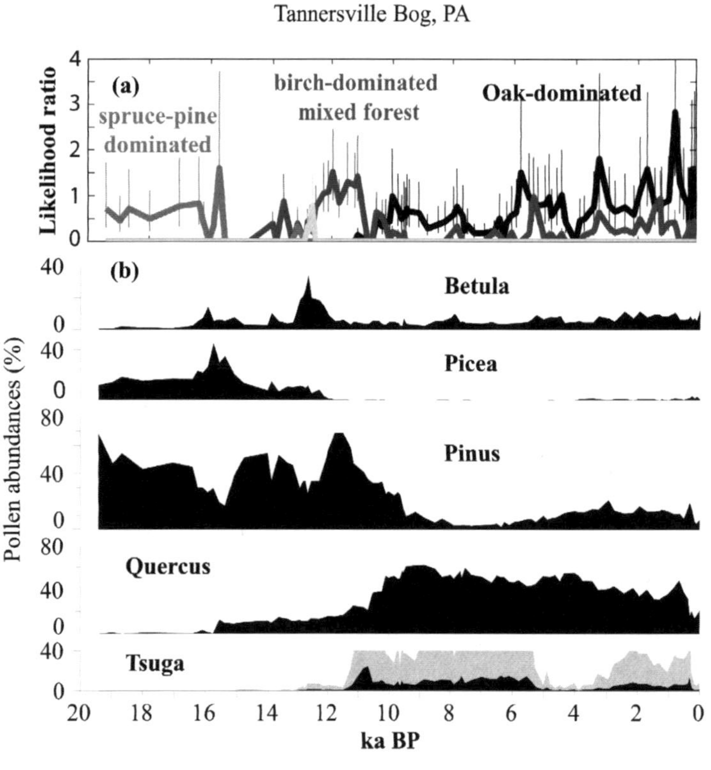

Figure 4. Panel (a) shows likelihood of analogs to modern ecosystems for Tannersville Bog since the Last Glacial Maximum. Colored lines indicate the likelihood of analog to modern ecosystem types. Red: spruce-pine dominated forest; blue: birch-dominated mixed forest; green: conifer woodland/savanna; black: oak-dominated forest. Vertical bars show 50% Bayesian confidence interval of likelihood estimates. Panel (b) shows pollen abundances of representative taxa at Tannersville Bog [33]. For *Tsuga* pollen, the grey area shows the abundance amplified by five times. The grey areas above the 40% upper bound are trimmed for the presentation of the figure.

3.2. Temporal and spatial patterns of antiquities of modern ecosystems

The temporal and spatial patterns of the origination of modern ecosystems are different among ecosystem types (Figure 5). The modern grassland ecosystem (Figure 5a) originated in the study area about 8 ka BP, and is mostly constrained to the western edge of the study region, including Minnesota, Iowa, southwest Wisconsin, and western Illinois. This position coincides with maximum eastern extent of the prairie-forest ecotone [34]. Younger sites in Ontario corresponded with recently developed oak savannas [35]. The modern spruce-dominated ecosystem (Figure 5b) originated around 6 ka BP in the southern Canadian Shield region. The spruce-dominated ecosystem appeared first in the south and then expanded to the north. The modern pine-dominated mixed forest ecosystem (Figure 5c) originated across the Great Lakes region around 10 ka BP. Sites showing later establishment of pine-dominated mixed forest showed an infilling pattern centered on the regions of initial establishment. The modern birch-dominated mixed forest (Figure 5d) originated around 10 ka BP in New England, southwestern Québec, and southeastern Ontario. Antiquities of pollen sites have three modes centered on early Holocene (~10 ka BP), mid-Holocene (~6 ka BP), and late Holocene (~3 ka BP). The mid-Holocene establishments are located northwest of the earliest sites, while the late Holocene establishments are located southeast of the earliest sites.

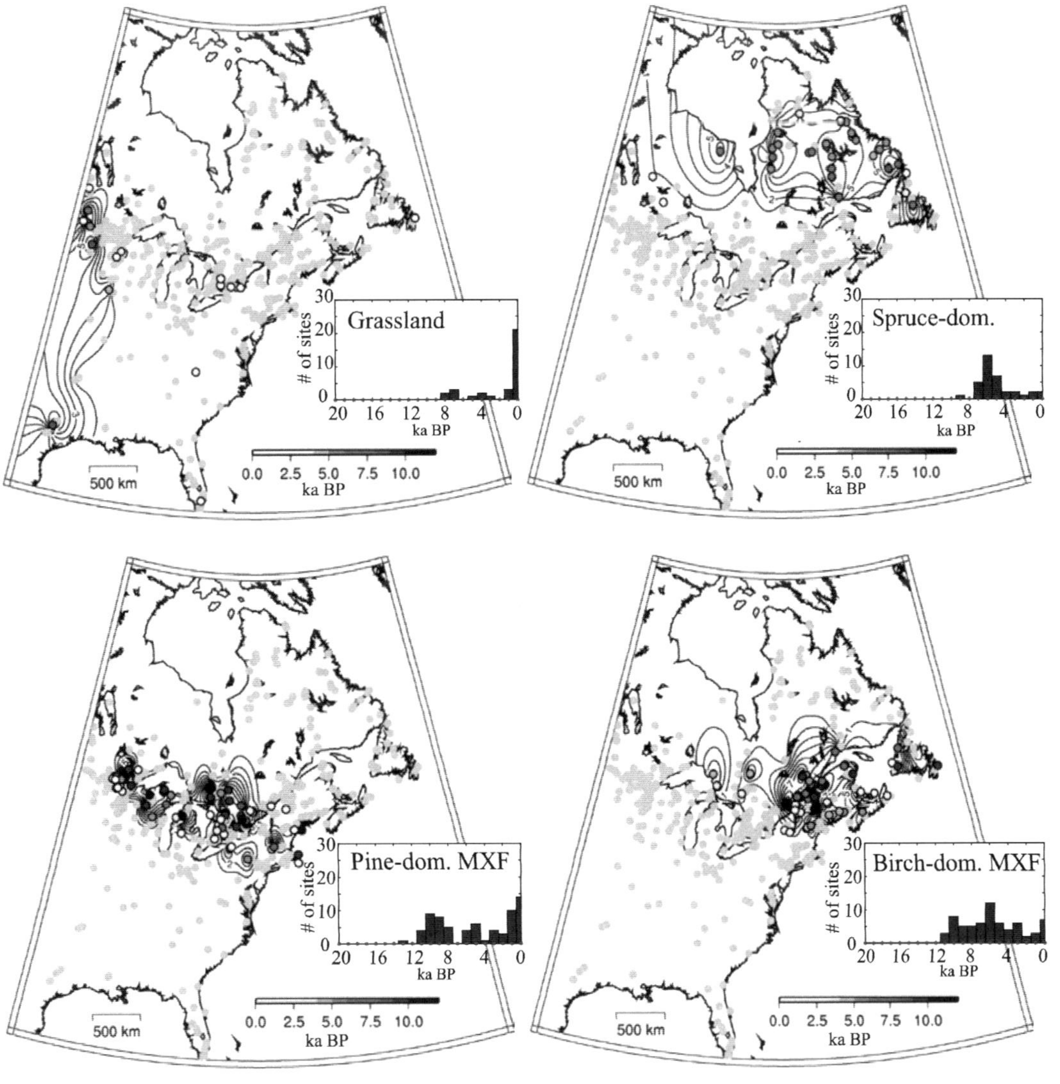

Figure 5. Examples of origination of modern ecosystems. a) Grassland ecosystem, b) Spruce-dominated ecosystem, c) Pine-dominated mixed forest ecosystem, d) Birch-dominated mixed forest ecosystem. The antiquity of a site is indicated by the color/grey scale of the dot. Contour lines connect sites of same antiquities. Grey dots represent pollen sites of other modern ecosystems.

Our analyses show that different ecosystems originated at different times, with different patterns and rates of expansion. Modern ecosystems at nearly all sites are younger than ~11.5 ka BP. A couple of exceptional sites with modern mixed forest ecosystems have dates of origination no older than13 ka BP. Within the Holocene, there was sufficient environmental change to drive ecosystem turnover in many regions.

4. Discussion and future directions

We discuss here the performance of our approaches in identifying and interpreting ecosystem antiquity and turnover using pollen data. We also identify challenges to be met with further studies.

Previous studies pointed out the optimal dissimilarity thresholds vary among different ecosystem types [17]. In our analyses, the ecosystem specific thresholds for the 15 ecosystem identified range from 0.06 to 0.23. Given that the objective of the work is to identify the antiquity and turnover for specific ecosystems, it is necessary to apply ecosystem specific dissimilarity thresholds, rather than a single dissimilarity threshold.

An ecosystem transition inferred from pollen records may be gradual rather than discrete, since landscapes are often heterogeneous due to factors such as mosaic of vegetation and compositional changes of taxa [36]. Furthermore, ecosystem transitions may be gradual, and ecotone boundaries and areal extent of individual elements of a vegetational mosaic may fluctuate with environmental fluctuation. To fully address the evolving nature of ecosystem transitions, therefore, likelihood estimates of modern analogs need to be assessed with uncertainties. Our Bayesian analyses showed large confidence intervals associated with likelihood ratio estimates, suggesting that a single estimate of expected value of the likelihood is not sufficient to make proper interpretations on ecosystem transition.

This continuous, as opposed to categorical, representation of ecosystem composition in the past better demonstrates the internal vegetation changes over time. In Tannersville Bog, the transition from birch-dominated mixed forest ecosystem to oak dominated ecosystem took place about 11 ka BP (Figure 4). The duration of the transition, shown by the dominance of likelihood-of-analog to the two ecosystems switching, lasted for more than 1k years. There are at least two potential explanations for the gradually evolving transition: 1) the birch-pine ecotone gradually shifted over the site as climate changed, or 2) the birch-dominated mixed forest patches persisted on the landscape throughout the transition and remained at low density after 11 ka BP. In summary, the likelihood representation could help to inform hypothesized potential mechanisms of vegetation change.

Challenging questions remain to be addressed with further analyses. For instance, in this study we used the modern pollen dataset to classify ecosystem types in this work. Then this classification was applied through time using modern analog technique. The underlying assumption here is that the variation within and among ecosystems in the past, is captured by the variation within and among spatially distributed modern ecosystems. Given the time scale of this particular study, however, the assumption needs more scrutiny. Interpretations on the rates of ecosystem change should also be made with the awareness of this assumption. An obvious limitation is that the ecosystem change within no-analog ecosystems (ecosystems have no modern analog) during the last deglaciation period cannot be represented by information about modern ecosystems [37]. Also, defining ecosystems by their pollen assemblages means that discriminations will depend on the spatial scale at which pollen assemblages represent vegetation [38], and on the taxonomic resolution of the pollen data [39]. Incorporation of plant-macrofossil data [39] into the analysis may result in finer discriminations (e.g., [2]). Finally, the number of ecosystems defined by modern pollen assemblages may also depend on the size of the region analyzed; had we used a smaller region (e.g., northeastern US), significant divisions might have emerged within our pine- and birch-dominated mixed forest categories, for instance.

We are currently using the approach in this study to assess the ecosystem turnovers in eastern North America since the last glacial maximum (LGM). Assessing properties associated with ecosystem turnover, including timing, duration, antecedent, and successive ecosystems can provide valuable climatic and ecological information. Are some ecosystems more persistent than others? Are there regions where ecosystems have been more persistent than others? With the approaches we have developed here, our future studies will be addressing these further questions.

Acknowledgments
This research was supported by National Science Foundation Grants DEB-0716951 and EPS-0447681. We thank an anonymous reviewer and Dave Mcwethy for thoughtful comments on the manuscript. PAGES provided funding for YL to attend 1st YSM and 3rd OSM to communicate this work.

References
[1] Williams J W, Shuman B N, Webb III T, Bartlein P J, Leduc P L 2004 *Ecological Monographs* **74** 309-334.
[2] Jackson S T 2006 *Journal of Vegetation Science* **17** 549-557
[3] Bosh J M and Hewlett J D 1982 *J. Hydrology* **55** 3-23
[4] Hu F S, Brubaker L B and Anderson PM 1993 *Can. J. Bot.* **71** 1133-1142
[5] Raich J W and Tufekciogul A 2000 *Biogeochemistry* **48** 71-90
[6] Post W M and Kwon K C 2000 *Global Change Biology* **6** 317-328
[7] Kutzbach J, Bonan G, Foley J and Harrison S P 1996 *Nature* **384** 623-626
[8] Pielke Sr R A, Marland G, Betts R A, Chase T N, Eastman J L, Niles J O, Niyogi D D S and Running S W 2002 *Philosophical Transactions: Mathematical, Physical and Engineering Sciences* **360** 1705-1719
[9] Brovkin V, Claussen M, Driesschaert E, Fichefet T, Kicklighter D, Loutre M-F, Matthews H D, Ramankutty N, Schaeffer M and Sokolov A 2006 *Climate Dynamics* **26** 587-600
[10] Bonan G B 2008 *Science* **320** 1444-1449
[11] Claussen M 2009 *Climate of the Past* **5** 203-216
[12] Suding K N and Hobbs R J 2009 *Trends in Ecology and Evolution* **24** 271-279
[13] Overpeck J T, Webb III T and Prentice IC 1985 *Quat. Res.* **23** 87-108
[14] Jackson S T, Webb R S, Anderson K H, Overpeck J T, Webb III T, Williams J W and Hansen B C S 2000 *Quaternary Science Reviews* **19** 489-508
[15] Williams J S, Shuman B N and Webb III T 2001 *Ecology* **82** 3346-3362
[16] Wahl E R 2004 *Review of Palaeobotany and Palynology* **128** 263-280
[17] Gavin D G, Oswald W W, Wahl E R and Williams J W 2003 *Quaternary Research* **60** 356-367
[18] Grimm E C 1987 *Computers & Geosciences* **13** 13-35
[19] Whitmore J, Gajewski K, Sawada M, Williams J W, Shuman B, Bartlein P J, Minckley T, Viau A E, Schafer S, Anderson and Brubaker L 2005 *Quaternary Science* **24** 1828-1848
[20] Williams J W, Shuman B, Bartlein P J, Whitmore J, Gajewski K, Sawada M, Minckley T, Shafer S, Viau A E, Webb III T, Anderson P M, Brubaker L B, Whitlock C, and Davis O K 2006. *An Atlas of Pollen-Vegetation-Climate Relationships for the United States and Canada*. American Association of Stratigraphic Palynologists Foundation, Dallas, TX, 293p.
[21] Chapin F S III, Matson P A and Mooney H A 2002 *Principles of terrestrial ecosystem ecology*. Springer-Verlag
[22] Lavorel S and Garnier E 2002 *Funct. Ecol.* **16** 545-556
[23] Chapin F S III 2003 *Ann. Bot.* **91** 455-463
[24] Diaz S, Hodgson J D, Thompson K, Cabido M, Cornelissen J H C 2004 *J. Veg. Sci.* **15** 295-304
[25] Bernabo J and Webb T 1977 *Quaternary Research* **8** 64-96
[26] Loveland T R, Reed B C, Brown J F, Ohlen D O, Zhu Z, Yang L and Merchant J W 2000 *Int. J. Remote Sensing* **21** 1303-1330
[27] Ter Braak C J F and Prentice I C 1988 *A theory of gradient analysis in Advances in Ecological Research* **18** 271-317
[28] Hartigan J A and Wong M A 1979 *Applied Statistics* **28** 100-108

[29] Brewer S, Alleaume S, Guiot J, Nicault A 2007 *Climate of the Past* **3** 355-366
[30] Bailey R G 1995 *Description of the Ecoregions of the United States* Forest Service Misc. Pub 1391 108p
[31] Eyre F H (ed.) 1980 *Forest Cover Types of the United States and Canada* Society of American Foresters 148p
[32] Strong W L and Hills L V 2005 *J. Biogeogr.* **32** 1043-1062
[33] Watts W A, 1979 *Ecological Monographs* **49** 427-469
[34] Williams J W, Shuman B, Bartlein P J 2009 *Global and Planetary Change* **66** 195-207
[35] Szeicz J M and Mcdonald G M 1991*Can. J. Bot.* **69** 1507-1519
[36] Pickett S T A and Cadenasso M L 1995 *Science* **269** 331-334
[37] Williams J W and Jackson S T 2007 *Frontiers in Ecology and the Environment* **5** 475-482
[38] Jackson S T 1994 Pages 253-286 in *Sedimentation of Organic Particles* (A. Traverse, editor). Cambridge Univ. Press.
[39] Jackson S T and Booth R K 2007 *Eos Trans. AGU* **88** GC31A-0111

Water availability reconstructions using tree-rings in the Valdivian rainforest ecoregion, Chile

R. Urrutia[ab], A. Lara[ab], M.P. Peña[a], D.A. Christie[a]

[a] Laboratorio de Dendrocronología, Instituto de Silvicultura, Facultad de Ciencias Forestales, Universidad Austral de Chile, Casilla 567, Valdivia, Chile; [b] Núcleo Milenio FORECOS, Iniciativa Científica Milenio and Fundación FORECOS

rociourrutia@uach.cl

Abstract. Water availability can be considered as one of the main restrictions for future development in South-Central Chile, due to the reported decreasing trends in precipitation in the last decades and the increasing demand for this resource. This issue makes the study of past water availability fundamental for the understanding of present and future variations. This paper presents a comparison of two water availability reconstructions within the Valdivian rainforest ecoregion (35°-48°S), one corresponding to a precipitation (37°-39.5° S) and the other to a streamflow reconstruction (41° S). This study shows that there are fundamental differences between them especially in the long term variability. However, there are also coincidences, mainly at higher frequency variations, such as at a bidecadal, decadal and annual scale. Another important finding is that these reconstructions show significant correlations with different climatic forcings in this area. The northern reconstruction presents a significant relationship with ENSO (El Niño Southern Oscillation), while the southern does the same with the AAO (Antarctic Oscillation Index).

1. Introduction

Water availability has been acquiring an increasing importance in South-Central Chile, especially considering the reported decreasing trends observed in the instrumental precipitation records during the last century (between 40% and 60% of decrease in the period 1901-2005, Trenberth et al., 2007), the unprecedented increment of extreme drought events during the last century in the context of the previous six (Christie et al. 2009), the regional glacier retreat (Masiokas et al. 2009), and the continuing decreasing rainfall projected for this area (Fuenzalida et al., 2007). This scenario imposes a restriction for future development even in areas with a relatively high amount of precipitation, such as the Valdivian rainforest ecoregion (35°-48° S, Lara et al., 2008). This region has been declared among the most threatened areas of the world by the Global Initiative 200 (WWF, World Bank), due to the high degree of plants and animals endemisms, as well as to the threats to different ecosystems and critical conservation status.

Given these circumstances and the development of important activities that require water within this region (agriculture, forestry, hydropower production), it is essential to understand changes in water availability in a long term perspective in this area.

Tree-rings provide annual records, being one of the proxies most commonly used to reconstruct past climate variability. Long streamflow and precipitation reconstructions can be developed by correlating tree-growth series with natural discharge and rainfall, respectively. Several streamflow reconstructions have been developed in North America (e.g., Stockton and Jacoby 1976; Meko et al. 1995; Hidalgo et al. 2000; Woodhouse 2001; Brito-Castillo et al. 2003; Woodhouse and Lukas, 2006; Meko et al., 2007) and a smaller number in South America, two of them in Argentina (Holmes et al. 1979 ; Cobos and Boninsegna 1983). In addition, a number of precipitation reconstructions have been developed in South America, especially in Chile and Argentina (Boninsegna, 1998, Villalba et al, 1998, Le Quesne et al., 2006).

As the number of studies reconstructing water availability is reduced in South America and two of them have already been developed in the Valdivian rainforest ecoregion in Chile, the objective of this study is to compare both reconstructions in order to assess differences in the past variability of water availability at these two distinct locations. One of them refers to a streamflow reconstruction (Lara et al., 2008) and the other to a precipitation reconstruction (Peña et al., not published), both covering the last 400 and 335 years, respectively.

2. Study area

The study area corresponds to the Valdivian rainforest ecoregion in South-Central Chile (35°-48° S). Climate in this area varies from Mediterranean in the north, where rainfall is concentrated in austral fall and winter (April-August, Miller, 1976) to Oceanic wet temperate in the south, with two dry months in summer (January-February). The precipitation reconstruction was developed for the Temperate-Mediterranean Transition zone in the Andes (TMTAN, approx. 37° - 39.5° S), corresponding to a transitional situation between southern temperate and northern Mediterranean influences. The streamflow reconstruction, on the other hand, was developed in a watershed characterized by an oceanic wet temperate climate with mild Mediterranean influence (41° S, Lara et al., 2008).

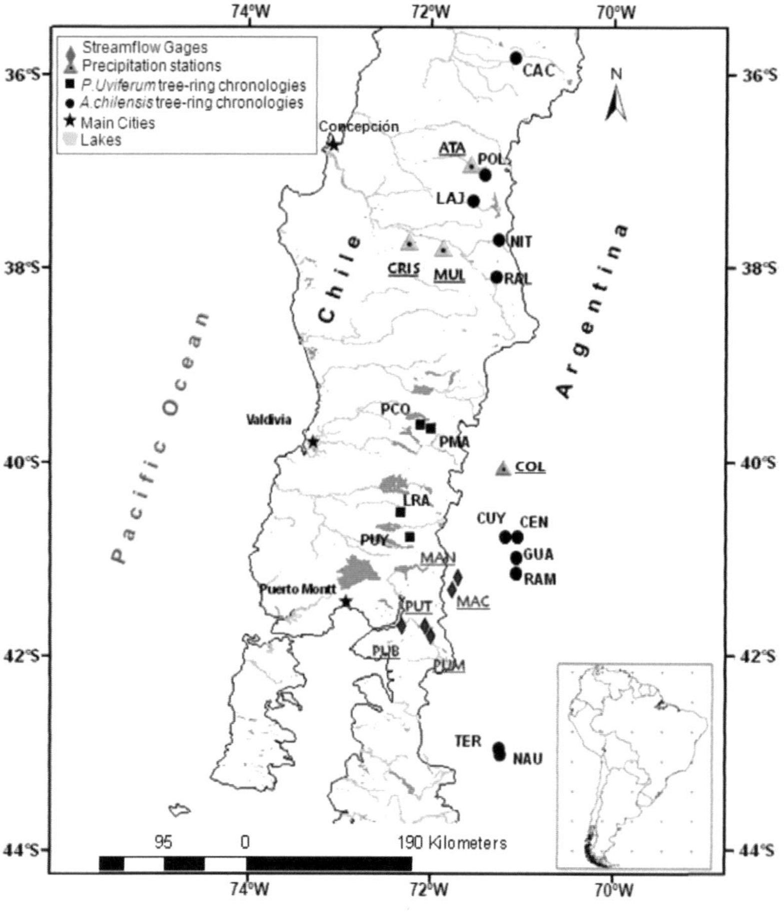

Figure 1. Regional map showing the location of the tree-ring chronologies, precipitation stations and streamflow gages that were used in the TMTAN precipitation reconstruction and the Puelo river streamflow reconstruction.

IOP Conf. Series: Earth and Environmental Science 9 (2010) 012006 doi:10.1088/1755-1315/9/1/012006

Climate variability in the ecoregion is driven by both low- and high latitude climatic forcings. Toward the northern portion (~35°-38°S), El Niño events are characterized by positive rainfall anomalies during late-spring (Montecinos & Aceituno, 2003). Conversely, toward the southern portion (40°-42°S), El Niño events are characterized by negative rainfall anomalies during summer (Montecinos & Aceituno, 2003).

The Antarctic Oscillation Index (AAO) also influences climate across the Valdivian rainforest ecoregion. The AAO corresponds to the leading principal component of 850hPa anomalies south of 20°S (Thompson and Wallace, 2000). Positive values in this index are associated with low geopotential height over Antarctica and increased values, and consequently dry conditions, over mid-latitudes and thus, over the study area.

3. Water availability reconstructions

A late spring-early summer precipitation reconstruction was developed for the Temperate-Mediterranean Transition zone of the Andes using five *Austrocedrus chilensis* tree-ring chronologies (CAC, POL, LAJ, NIT, RAL, Peña et al., not published, Figure 1). In order to preserve a large percentage of the low-frequency variance (decadal-multidecadal variability), the standardization process was done by fitting each ring width to a negative exponential or linear regression curve, using the TURBO ARSTAN program (Cook, 1985). This 335-year reconstruction was developed by regressing the November-December precipitation on principal components (PCs) of standard chronologies' selected lags (t-1, t, t+1, predictors) using a stepwise multiple regression. The reconstruction explained 41% of the total variance in the late spring-early summer precipitation for the whole calibration period (1937-2002, Figure 2). The selected reconstruction period represents 7.5% of the annual rainfall, representing a portion of the precipitation falling during the dry season.

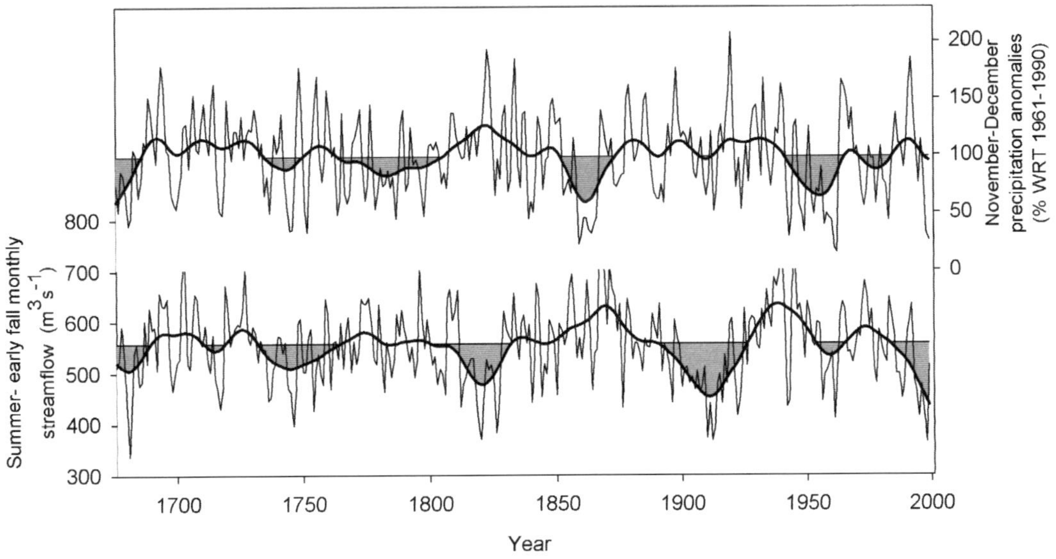

Figure 2. November-December precipitation reconstruction for the TMT of the Andes (top). December-May streamflow reconstruction for the Puelo river (bottom). Both reconstructions cover the period 1676-1999. To emphasize the long-term variations, a cubic spline version designed to reduce 50% of the variance in a sine wave with a periodicity of 25 years is also shown for both reconstructions (Cook and Peters, 1981).

Additionally, a summer-early fall streamflow reconstruction was developed for the Puelo river (Lara et al., 2008), a watershed shared with Argentina, that has its mouth at approximately 41° 35' S in the Reloncaví estuary (Figure 1). The 400-year streamflow reconstruction used three composite *Austrocedrus chilensis* and one composite *Pilgerodendrum uviferum* tree-ring chronologies which were also standardized using the method mentioned above. The reconstruction equation was estimated by regressing the summer-early fall streamflow on principal components of the four composite tree-ring chronologies. This reconstruction also accounted for low frequency variability, explaining 42% of the total variance in the December-May streamflow for the entire calibration period (1943-1999, Figure 2). The summer-early fall reconstruction period is the season when water availability becomes critical for different uses (salmon farming and sport fishing, among others) and represents 21% of the annual streamflow.

It is important to mention that the conifers used in this reconstruction are endemic of Chile and Argentina, andhave clear annual growth rings that are sensitive to moisture supply.

4. Comparison of both water availability reconstructions

Comparing both reconstructions (Figure 2), it is possible to see that they mostly present a different behaviour. The pronounced decreasing trend in the last decades in the streamflow reconstruction (also present in the instrumental record) is not observed in the precipitation reconstruction toward the north. Additionally, most of the multi-year periods of low and high water availability do not match between them, except for a few cases where there is a partial coincidence (e.g. around 1680, 1740, 1960). Despite this differential behaviour at a multi-year scale, there is a significant correlation between them (r=0.24, p<0.05).

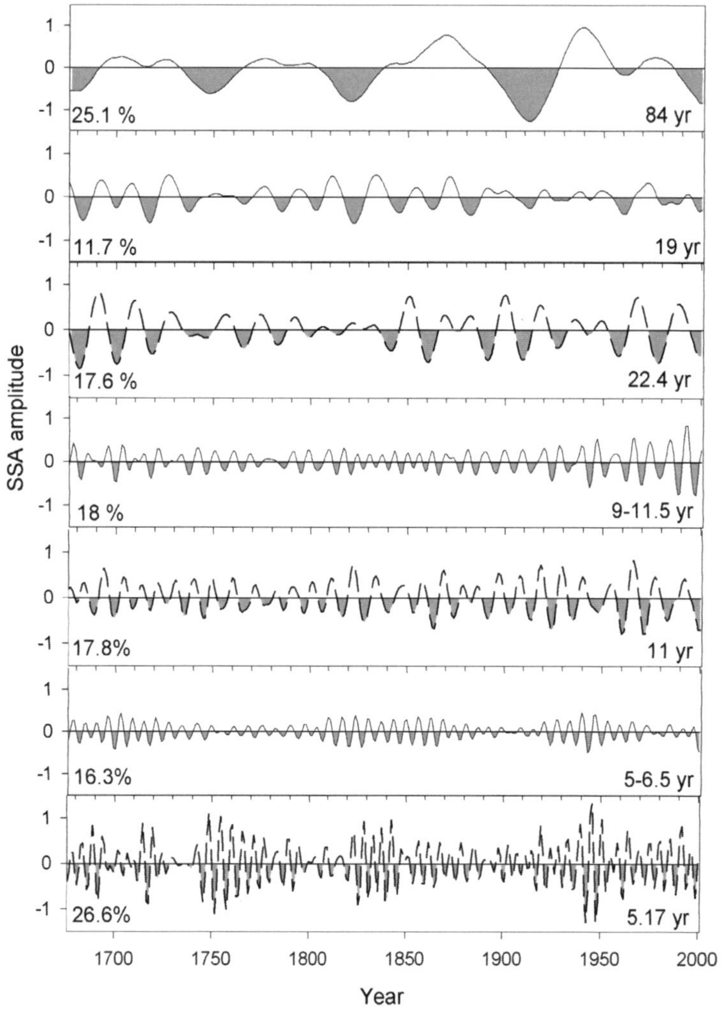

Figure 3. The most significant oscillation modes estimated using Singular Spectrum Analysis (units are dimensionless) for both reconstructions (solid lines for the Puelo river streamflow reconstruction and dashed lines for the TMTAN precipitation reconstruction). Periods in years and the percentage of variance associated with each waveform are shown in the lower right and left corners of the streamflow waveforms, respectively.

The major differences between both reconstructions might be explained by looking at the main oscillation modes, obtained through singular spectral analyses using a lag of 12% of the series length (31 years,Vautard and Ghil, 1989, Figure 3). The main difference is that the southern reconstruction shows a prominent multidecadal cycle of 84 years that explains the largest percent of the variance (25%), and that is not observed in the precipitation reconstruction toward the north. It has been stated that the decreasing trend of the Puelo river in the last six decades forms part of this cycle that has been increasing its negative amplitude through time (Lara et al., 2008). In the case of the precipitation reconstruction the largest percent of the variance is explained by a much shorter-term mode, the 5.17 year cycle. However, it is possible to observe that there are coincidences in the bidecadal cycles of 19 and 22 years in the precipitation and streamflow reconstructions, respectively, and also in the decadal cycle of around 11 years presented in both cases. Correlations among them are r= 0.4 (n=325) for the bidecadal cycle and r=0.27 (n=325), for the decadal cycle. These correlations do not account for autocorrelation in the series, therefore the level of significance is not provided.

The bidecadal and decadal oscillation modes may be explained by the 22-year Hale cycle and the 11-year sunspot cycles, respectively, both invoked to explain these modes, not only in the Puelo river streamflow reconstruction, but also in two millenial temperature reconstructions using *Fitzroya cupressoides* (Villalba et al., 1996, Lara et al., 2008).

Finally, both reconstructions present the oscillation mode close to 5 years, although the amplitude is much larger in the case of the precipitation reconstruction. This cycle might be attributed to the ENSO activity that is more conspicuous to the north.

5. Water availability reconstructions and climatic forcings

As mentioned before, ENSO and the AAO have a strong influence on climatic variability in this area. Figure 4 shows the best correlations found between water availability reconstructions and these forcings. In this figure it is possible to see that the northern late spring-early summer precipitation reconstruction is significantly and negatively correlated with the spring-early summer Southern Oscillation Index (SOI, http://www.cru.uea.ac.uk/cru/data/soi.htm). This means that high (low) SOI values or La Niña (El Niño) events are related with low (high) precipitation anomalies in this area. Conversely, the summer-early fall streamflow reconstruction is mostly related to the Antarctic Oscillation Index (http://jisao.washington.edu/data/aao/slp/) in summer, relationship also found for the observed streamflow record (Lara et al., 2008). Although the northern reconstruction includes part of the summer, no significant correlation was found between this one and the AAO and between the southern reconstruction and the SOI. These correlations demonstrate the influence of climatic forcings at different seasons of the year on the variability of water availability in this area. Toward the south the recorded positive trend of the AAO (Marshall, 2003) could be influencing the pronounced decreasing trend in the streamflow in this area, however its influence is not observed toward the north, at least during the late spring-early summer season.

6. Conclusions

Reconstructing past water availability using tree-rings in south-central Chile is essential to understand the last decades' changes in a long term perspective and to improve the future water resources allocation in the Valdivian rainforest ecoregion in Chile. This region is especially important due to its conservation status and the different activities that develop and compete within this area. In this study, two existing water availability reconstructions were compared, providing a broad characterization of the low frequency water availability changes that have affected this ecoregion in the last 325 years. Their comparison also contributes to the understanding of the main differences in variability at two locations within the area (approx. 37°-39.5° and 41° S).

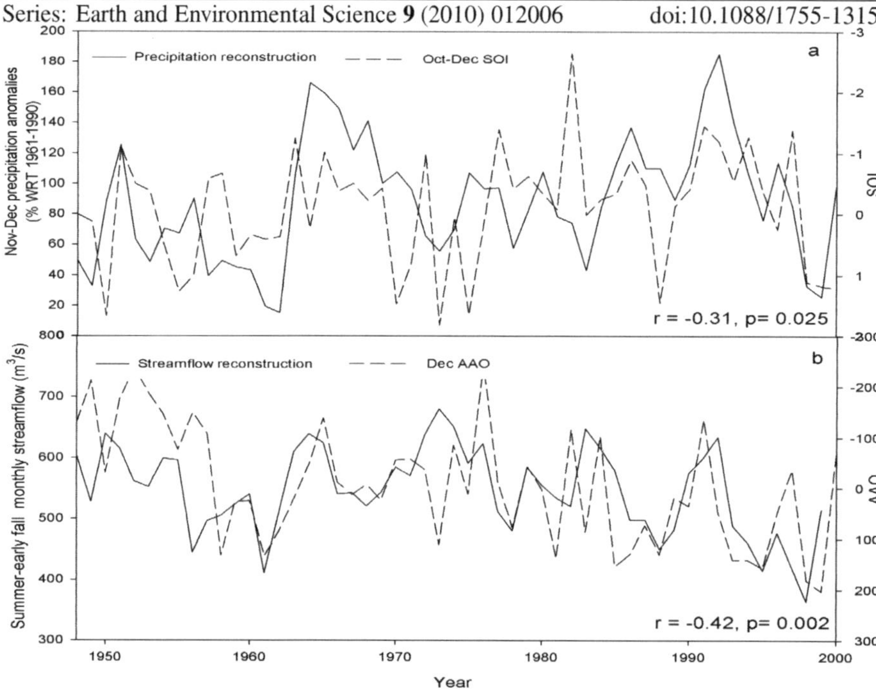

Figure 4. a) Late spring-early summer (November-December) precipitation reconstruction for the TMTAN zone and its relation with the October-December Southern Oscillation Index (SOI) for the 1948-1999 period. This index is inverted to clearly appreciate wet (El Niño) and dry (La Niña) periods. Correlation between both is shown. b) Summer-early fall (December-May) Puelo streamflow reconstruction and its relation with the December Antarctic Oscillation Index for the same period. This index has been inverted to facilitate the comparison with the streamflow data. Correlation between both is also shown.

Both the precipitation reconstruction in the northern site and the streamflow reconstruction in the southern area present major differences in the multidecadal low-frequency variability that characterize them, but not in the decadal-bidecadal variability that is more closely related. This could be partially attributed to the differential and contrasting effects of ENSO events between both locations and to the stronger modulation of AAO over the local hydroclimate toward the southern portion of the study area (Garreaud et al. 2009). However, it is important to be careful in the comparison, since streamflows may carry the effects of other factors besides precipitation, and since hydrological events are usually out of phase with or lag the occurrence of meteorological episodes.

Both reconstructions appear to be affected by climatic forcings in this area, ENSO especially influencing late spring-early summer precipitation behavior toward the north and the AAO affecting summer-early fall water availability toward the south. The current and projected increasing trend in the AAO, could be a real threat for water availability in the ecoregion, enhancing the actual decreasing trend in streamflow in the southern area, although its influence might be less prominent toward the north. More studies are needed to comprehend latitudinal variations in past water availability in South-Central Chile, as well as the forcings that are responsible for these changes.

Acknowledgements

We thank the CRN-II 2047 project of the Inter American Institute for Global Change Research (IAI) and PAGES for the financial support given to R. Urrutia and D. Christie to attend the First YSM and the Third OSM in Corvallis, July 2009. We also thank to the Fondecyt grant N°1090479 and the FORECOS Scientific Nucleus.

References

[1] Boninsegna JA. (1988) Santiago de Chile winter rainfall since 1220 as being reconstructed by tree rings. Quaternary South American Antarctic Peninsula 6:67–87.

[2] Briffa KR.(1995) Interpreting high-resolution proxy climate data: the example of dendroclimatology. In: Von Storch H, Navarra A (eds) Analysis of climate variability: applications of statistical techniques. Springer, Berlin, pp 77–94.

[3] Brito-Castillo L, Díaz-Castro S, Salinas-Zavala CA, Douglas AV. (2003) Reconstruction of long-term winter streamflow in the Gulf of California continental watershed. Journal of Hydrology 278:39–50.

[4] Christie, D.A., Boninsegna, J.A., Cleaveland, M.K., Lara, A., LeQuesne, C., Morales, M.S., Mudelsee, M., Stahle, D.W. and Villalba, R. (2009) Aridity changes in the Temperate-Mediterranean transition of the Andes since AD 1346 reconstructed from tree-rings. Climate Dynamics doi:10.1007/s00382-009-0723-4.

[5] Cobos DR and JA. Boninsegna (1983) Fluctuations of some glaciers in the upper Atuel River basin, Mendoza, Argentina. Quaternary South American Antarctic Peninsula 1:61–82.

[6] Cook ER. (1985) A time series analysis approach to tree ring standardization. Ph.D. dissertation. University of Arizona, Tucson.

[7] Cook ER and K. Peters (1981) The smoothing spline: a new approach to standardizing forest interior ring-width series for dendroclimatic studies. Tree-Ring Bulletin 41:45–53.

[8] Holmes RL, Stockton CW, LaMarche VC. (1979) Extension of river flow records in Argentina from long tree-ring chronologies. Water-Resources Bulletin 15(4):1081–1085.

[9] Jain S, Woodhouse CA, Hoerling MP. (2002) Multidecadal streamflow regimes in the interior Western United States: implications for the vulnerability of water resources. Geophysical Research Letters 29(21).

[10] Lara, A., Villalba, R., and Urrutia, R. (2008). A 400-year tree-ring record of the Puelo River summer–fall streamflow in the Valdivian rainforest eco-region, Chile. Climatic Change, doi 10.1007/s10584-007-9287-7. Vol 86 (3-4): 331-356.

[11] Le Quesne, C., Stahle, D., Cleaveland, M., Therrell, M, Aravena, J., and Barichivich. J. (2006) Ancient Austrocedrus Tree-Ring Chronologies Used to Reconstruct Central Chile Precipitation Variability from A.D. 1200 to 2000. Journal of Climate 19: 5731-5744.

[12] Garreaud, R., Vuille, M., Compagnucci, R. and Marengo, J. (2009) Present-day South American climate. Palaeogeography, Palaeoclimatology, Palaeoecology 281: 180–195.

[13] Marshall, G. (2003) Trends in the Southern Annular Mode from observations and reanalyses. Journal of Climate 16: 4134-4143.

[14] Masiokas, M.H., Rivera, A., Espizua, L.E., Villalba, R., Delgado, S. and Aravena, J.C. (2009) Glacier fluctuations in extratropical South America during the past 1000 years. Palaeogeography, Palaeoclimatology, Palaeoecology 281: 242–268.

[15] Meko, D., Stockton, C.W. and Boggess, W.R. (1995) The tree-ring record of severe sustained drought. Water Resources Bulletin 31 (5): 789-801.

[16] Meko D. M., Woodhouse C. A., Baisan C. H., Knight T., Lukas J. J., Hughes M. K. and Salzer M. W. (2007) Medieval drought in the Upper Colorado River Basin. Geophysical Research Letters 34(10, L10705), 10.1029/2007GL029988.

[17] Miller, A. (1976) The Climate of Chile. In: Schwerdtfeger, W. (ed.). World Survey of Climatology. Climates of Central and South America. Elsevier, Amsterdam, The Netherlands, pp. 113-131.

[18] Pederson, N, Jacoby, G.C, D'Arrigo, R.D., Cook, E.R., Buckley, B.M. Dugarjav, C. and Mijiddorj, R. (2001) Hydrometeorological reconstructions for northeastern Mongolia derived from tree rings: 1651-1995. Journal of Climate 14 (5): 872-881.

[19] Rutlant., J and H. Fuenzalida (1991). Synoptic aspects of the Central Chile rainfall variability associated with the southern oscillation. International Journal of Climatology 11:63-76.

[20] Stockton, C.W and GC. Jr. Jacoby (1976) Long-term surface-water supply and streamflow

trends in the Upper Colorado River Basin based on tree-ring analysis, Lake Powell Res. Project Bull. 18, Inst. of Geophys. and Planet. Phys., Univ. of Calif. Los Angeles.

[21] Trenberth, K.E., P.D. Jones, P. Ambenje, R. Bojariu, D. Easterling, A. Klein Tank, D. Parker, F. Rahimzadeh, J.A. Renwick, M. Rusticucci,B. Soden and Zhai, P. (2007) Observations: Surface and Atmospheric Climate Change. In: Climate Change 2007: The Physical Science Basis. Contribution of Working Group I to the Fourth Assessment Report of the Intergovernmental Panel on Climate Change [Solomon, S.,D. Qin, M. Manning, Z. Chen, M. Marquis, K.B. Averyt, M. Tignor and H.L. Miller (eds.)]. Cambridge University Press, Cambridge, United Kingdom and New York, NY, USA.

[22] Urrutia, R., A. Lara, R. Villalba, C. Le Quesne and E. Cuq. Multi-century tree-ring reconstruction of annual streamflow for the Maule watershed, South-Central Chile (Submitted to Journal of Hydrology).

[23] Vautard, R. and M. Ghil (1989) Singular Spectrum Analysis in Nonlinear Dynamics, with Applications to Paleoclimatic Time Series, Physica D 35: 395–424.

[24] Villalba R., Boninsegna, J.A., Lara, A., Veblen, T., Roig, F., Aravena, J.C. and Ripalta, A. (1996) Interdecadal climatic variations in millennial temperature reconstructions from southern South America. In: Jones PD, RS Bradley and J Jouzer (eds.) Climatic Variations and Forcing Mechanisms of the Last 2000 years. NATO ASI Series 141, Springer- Verlag Berlin-Heidelberg, Berlin, pp. 161-189.

[25] Villalba, R., Cook, E., Jacoby, G., D'Arrigo, R., Veblen, T. and Jones, P. (1998) Tree-ring based reconstructions of northern Patagonia precipitation since AD 1600. The Holocene 8 (6): 659-674.

[26] Woodhouse, C.A. (2001) A tree-ring reconstruction of streamflow for the Colorado Front Range. Journal of American Water Research Association 37 (3): 561-569.

[27] Woodhouse, C.A. and J. Lukas (2006) Multi-Century Tree-Ring Reconstructions of Colorado Streamflow for Water Resource Planning. Climatic Change 78 (2-4):293-315.

How different proxies record precipitation variability over southeastern South America

Cristiano M Chiessi[1,2], Stefan Mulitza[1], Jürgen Pätzold[1] and Gerold Wefer[1]

[1]MARUM-Center for Marine Environmental Sciences, University of Bremen, Leobener Straβe, 28359 Bremen, Germany

[2]Now at Center for Earth System Science, National Institute for Space Research, Av. dos Astronautas 1758, CEP12227-010 São José dos Campos SP, Brazil

E-mail: chiessi@uni-bremen.de

Abstract. Detrending natural and anthropogenic components of climate variability is arguably an issue of utmost importance to society. To accomplish this issue, one must rely on a comprehensive understanding of the natural variability of the climate system on a regional level. Here we explore how different proxies (e.g., stalagmite oxygen isotopic composition, pollen percentages, bulk sediment elemental ratios) record Holocene precipitation variability over southeastern South America. We found a general good agreement between the different records both on orbital and centennial time-scales. Dry mid Holocene, and wet late Holocene, Younger Dryas and a period between ~9.4 and 8.12 cal kyr BP seem to be pervasive features. Moreover, we show that proxy-specific sensitivity can greatly improve past precipitation reconstructions.

1. Introduction

Southeastern South America (SESA) is the most densely populated, industrialized and agriculturally productive region in the continent. Domestic, industrial and agricultural water and power supply strongly relies on seasonal precipitation associated with the South American summer monsoon (SASM). Thus, hydrological changes are of great concern in the context of anthropogenic climate change. Assessing past changes in precipitation has the potential to clarify the mechanisms that govern precipitation under different boundary conditions. Here we present bulk sediment elemental ratios (Al/Si) reflecting changes in Holocene chemical weathering in SESA and compare this record with charcoal accumulation rates, stalagmite oxygen isotopic composition ($\delta^{18}O$), and pollen percentages in order to explore how different proxies record precipitation variability over SESA.

2. Environmental setting and methods

2.1. Environmental setting

Precipitation over SESA is mainly related to the southward expansion and intensification of the SASM, while austral winter rainfall associated with mid-latitude cyclonic activity over the South Atlantic plays a secondary role (figure 1) [1, 2]. During austral summer, Amazon moisture is transported southwards by the South American low level jet, producing strong precipitation in the South Atlantic convergence zone (SACZ) [3]. During austral winter, incursions of mid-latitude air masses into the southern SESA associated with episodes of enhanced cyclonic activity over the South Atlantic generate winter rainfall that progress northwards along the Atlantic coast [2]. Interannual variability in precipitation over SESA has been related to El Niño-Southern Oscillation (i.e., El Niño is associated with above-normal precipitation) whereas interdecadal changes were associated to sea surface temperature (SST) fluctuations in the western South Atlantic (i.e., positive SST anomalies were associated with above-normal precipitation) [4].

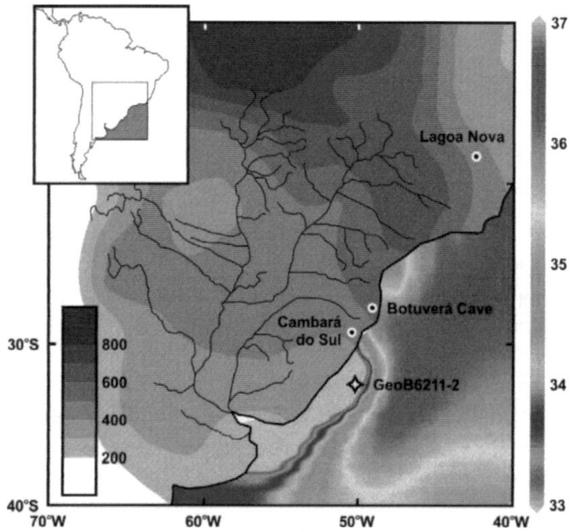

Figure 1. Location of sites GeoB6211-2 (this study) in the western South Atlantic, and Lagoa Nova [12], Botuverá Cave [14, 15] and Cambará do Sul [18] in southeastern South America. Color shading shows long-term mean annual sea surface salinity (in psu) [27], and grey shading shows long-term mean December-February precipitation (in mm) [28]. The main tributaries of the La Plata River drainage basin are also shown.

2.2. Methods
Our Al/Si record was obtained from marine sediment core GeoB6211-2 (32.50°S, 50.24°W, 657 m water depth) (figure 1). We focused on the last 14 cal kyr BP where we found relatively constant sedimentation rates (~9cm kyr^{-1}). Details of GeoB6211-2 calibrated ^{14}C-based age model are provided in Chiessi et al. [5]. Aluminum and Si intensities were determined in bulk sediment using X-ray fluorescence (XRF) core scanner. Prior and after analysis, the instrument was calibrated against a set of pressed powder standards. Further methodological details can be found in Chiessi et al. [6].

Considering that the eolian input [7] and the biogenic opal accumulation rate [8] in our core site is relatively small, we interpret the Al/Si ratio in GeoB6211-2 as an indicator of the intensity of chemical weathering in the La Plata River drainage basin (PRDB) (figure 1), related to precipitation variability over the basin. The PRDB is the fifth largest river basin in the world, extending over 3.1 million km^2, and draining a significant portion of the SASM precipitation. Depetris et al. [9] and Depetris and Pasquini [10] showed that the Al/Si ratio in the suspended sediment load of the PRDB northwestern tributaries is significantly lower than the same ratio for the northeastern tributaries. The authors associated this difference to the less (more) intense chemical weathering conditions in the northwestern (northeastern) sector of the basin, largely controlled by the present-day distribution in precipitation.

3. Results

Our Al/Si record shows centennial-scale fluctuations superimposed on a long-term trend of decreasing values (0.03) from 14 until ~6.25 cal kyr BP, and increasing values (0.07) from 6.25 cal kyr BP until the present (figure 2e). Maximum and minimum values are 0.32 and 0.21, respectively. A peak with relatively low amplitude is observed between 12.62 and 11.52 cal kyr BP, whereas a peak with relatively high amplitude appears between ~9.4 and 8.12 cal kyr BP. Two discrete increases in the Al/Si ratio are present around 4.52 and 2.36 cal kyr BP.

4. Discussion

Charcoal accumulation rates in continental sediments can be considered an index of the overall fire frequency, intensity and extent of the regional fire regime [11]. While the incidence of fire over space and time is influenced by complex interactions between climate, fuels, and ignition, the Lagoa Nova record (figures 1 and 2a) was interpreted to mainly reflect climate conditions around the sampling site [12]. Fire frequency was relatively high between ~11 and 9.5 cal kyr BP, and between ~8.5 and 7 cal kyr BP. From ~7 cal kyr BP towards the present, fire frequency decreased gradually. Together with palynological evidence from the same site, the author assigned a long dry season of about 6 months and annual precipitation lower than today for the two periods of high fire frequency. On the other hand, the intermediate period of low fire frequency as well as the mid and late Holocene showed wetter climatic conditions with an annual dry season of about 4 months.

The variability in $\delta^{18}O$ of tropical and subtropical stalagmites depends on temperature, the amount effect, and the moisture source [13]. Oxygen isotopic composition of Botuverá Cave stalagmites (figures 1 and 2b) has been mainly related to changes in the moisture source [14, 15]. Early and mid Holocene $\delta^{18}O$ values are relatively high and were associated to South Atlantic extratropical moisture, whereas late Holocene low $\delta^{18}O$ values are typical of Amazon-derived moisture. A sharp peak characterized by low values was observed to coincide with the Younger Dryas (YD) (~12.85-11.7 cal kyr BP [16]). Thus, the moisture source feeding precipitation over Botuverá Cave closely followed changes in austral summer insolation at subtropical latitudes (figure 2f) but also varied on shorter time-scales reflecting abrupt shifts in atmospheric circulation related to changes in the Atlantic meridional overturning circulation (AMOC) [14, 15].

Some plant taxa are particularly sensitive to climate conditions, and in the case of good preservation of their palynomorphs may constitute important proxies for past climate. A good example is the tree *Araucaria angustifolia*, typical of southern Brazilian forests, that requires a minimum of 1400 mm annual rainfall without a marked annual dry season [17]. Moreover, shifts in grassland/forest boundaries in SESA are an important tool for reconstructing past climates [17]. The herb family Poaceae is one of the main representatives of SESA grasslands and temporal changes in its relative abundance in site Cambará do Sul (figure 1) are depicted in figure 2c together with temporal changes in the relative abundance of *A. angustifolia* [18]. Increasing percentages of Poaceae through the early and mid Holocene together with constant values of *A. angustifolia* suggest a trend towards drier conditions with a long annual dry season. After ~5 cal kyr BP, an increase in *A. angustifolia* and a decrease in Poaceae abundances points to a gradual humidification until ~1 cal kyr BP when the forest biome replaced grasslands, reflecting the onset of the wettest period in the last 14 cal kyr BP with no marked annual dry season.

Long-term changes in chemical weathering in the PRDB as recorded by GeoB6211-2 (figure 1) Al/Si ratios (figure 2e) followed subtropical austral summer insolation (figure 2f). A similar orbital-scale trend is observed in Botuverá Cave $\delta^{18}O$ values and in Cambará do Sul Poaceae abundances. The common signal corroborate model simulations pointing to a decrease in moisture transport from the Amazon basin to SESA, and a reduced intensity of the SACZ during the mid Holocene in relation to present-day climate [19]. Low (high) peak austral summer insolation during the mid Holocene (late glacial and late Holocene) would be responsible for the observed pattern by shifting the mean position of the Intertropical Convergence Zone (ITCZ) to the north (south) and affecting the SASM circulation (e.g., South American low level jet, SACZ) [14]. Because the relative abundances of *A. angustifolia* do not show high values for the late glacial and the early Holocene, we assume the occurrence of a relatively long annual dry season despite of the relatively high total annual rainfall for that period.

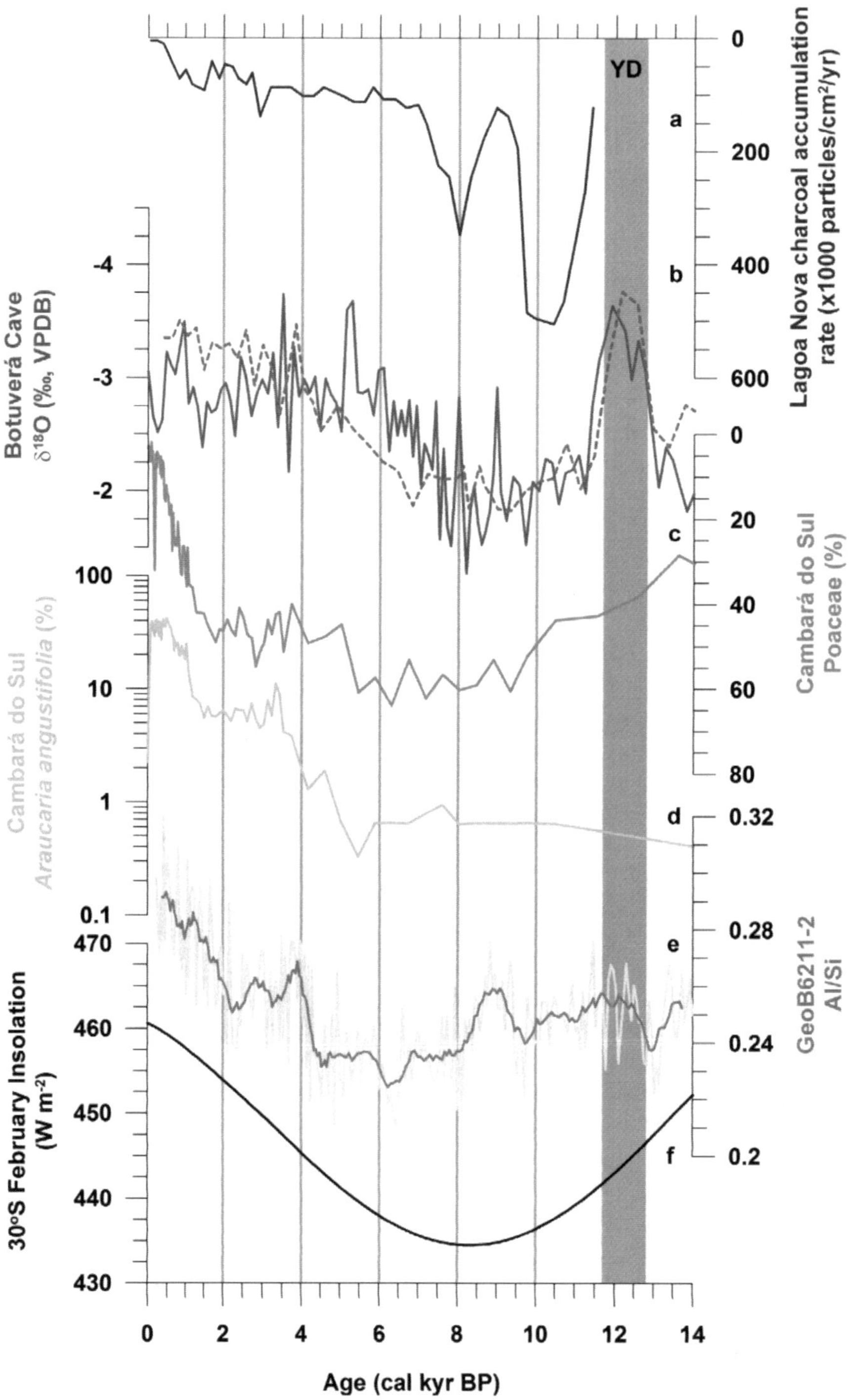

Figure 2. Proxy records of precipitation variability over southeastern South America compared to subtropical austral summer insolation. **a** Lagoa Nova [12], **b** Botuverá Cave (bold line Cruz et al. [14] with age model of Cruz et al. [29], dashed line Wang et al. [15]), **c** and **d** Cambará do Sul [18], **e** GeoB6211-2 raw data (pale green line) and 9 point moving average (green line) (this study), and **f** Berger and Loutre [30]. All proxy curves show more humid conditions towards the top of the panel. Note the reverse scale in **a**, **b** and **c**. The grey bar shows the duration of the Younger Dryas (YD) [16].

Centennial-scale changes in the Al/Si record are also present in other paleoprecipitation records from SESA when the necessary temporal resolution is available. The low amplitude peak around 12 cal kyr BP has a counterpart in the Botuverá Cave $\delta^{18}O$ record, while the much stronger peak between ~9.4 and 8.12 cal kyr BP is clearly related to a decrease in charcoal accumulation rate in the Lagoa Nova record. A major slowdown event in AMOC strength happened during the YD [20] and was claimed to affect atmospheric circulation in SESA [14, 15]. As summarized by Chiessi et al. [6], periods of weak AMOC are related to anomalously warm SSTs in the South Atlantic [21, 22] and increased SACZ activity, displacing the main belt of SASM precipitation to the south. Amplified SACZ activity would increase rainfall over the PRDB, the intensity of chemical weathering in the basin and the Al/Si ratio of the basin's suspended sediment load. Another mechanism linking AMOC strength and SASM/SACZ involves the position of the ITCZ. Modeling results show that during periods of weak AMOC the ITCZ is shifted to the south causing a positive anomaly in moisture transport into the Amazon basin [23]. The equatorial anomalies would enhance convection in the Amazon basin, reinforcing moisture transport towards SESA during austral summer (i.e., SASM/SACZ) [3].

The peak between ~9.4 and 8.12 cal kyr BP is too long to be related to the sudden 8.2 kyr event, but could be indeed associated to a multi-century climate change reported in many monsoon areas around the globe [24]. In this case, it would be part of a repeating pattern of longer-term anomalies during the Holocene, with its most recent manifestation during the Little Ice Age. It is noteworthy that the peak of high precipitation in the PRDB between ~9.4 and 8.12 cal kyr BP is remarkably synchronous to high levels in Lake Titicaca [25] and to an early peak in human occupation in central Brazil [26].

5. Conclusions

We found a general good agreement between different proxies (e.g., stalagmite $\delta^{18}O$, pollen percentages, and bulk sediment elemental ratios) recording precipitation variability over SESA. While most of them show an austral summer insolation-driven long-term variability with a dry mid Holocene and a wet late Holocene, records with high temporal resolution depict centennial-scale changes that seems to be pervasive (e.g., a wet YD). We conclude that proxy-specific sensitivity (e.g., the ability to distinguish different moisture sources in stalagmite $\delta^{18}O$, the duration of the dry season in *A. angustifolia*, and the basin integrated precipitation in bulk sediment elemental ratios) should be used in order to draw a more complete scenario of paleoprecipitation variability.

Acknowledgments
We thank U. Röhl for help with the XRF analyses and reviewers P.G.C. Amaral and H. Behling for constructive comments. This study was funded by the DFG-Research Center/Excellence Cluster "The Oceans in the Earth System", and the CNPq-Brazil Fellowship granted to C.M. Chiessi. We are grateful to the US NSF and PAGES for providing financial support to C.M. Chiessi to attend the PAGES 1st Young Scientists Meeting. Data presented in this study are available at the Pangaea database (http://www.pangaea.de).

References

[1] Zhou J and Lau K –M 1998 *J. Clim.* **11** 1020
[2] Vera C S, Vigliarolo P K and Berbery E H 2002*Mon. Weather Rev.* **130** 684
[3] Marengo J A, Soares W R, Saulo C and Nicolini M 2004 *J. Clim.* **17** 2261
[4] Robertson A W and Mechoso C R 2000 *Mon. Weather Rev.* **128** 2947
[5] Chiessi C M, Mulitza S, Paul A, Pätzold J, Groeneveld J and Wefer G 2008 *Geology* **36** 919
[6] Chiessi C M, Mulitza S, Pätzold J and Wefer G 2009 *Geophys. Res. Lett.* **36** doi:10.1029/2009GL039914
[7] Mahowald N M, Muhs D R, Levis S, Rasch P J, Yoshioka M, Zender C S and Luo C 2006 *J. Geophys. Res.* **111** doi:10.1029/2005JD006653

[8] Romero O and Hensen C 2002 *Mar Geol* **186** 263
[9] Depetris P J, Probst J -L, Pasquini A I and Gaiero D M 2003 *Hydrol. Process.* **17** 1267
[10] Depetris P J and Pasquini A I 2007 *The Middle Paraná River: Limnology of a Subtropical Wetland* ed M H Iriondo et al. (Berlin, Springer-Verlag) p 143
[11] Power M J et al. 2008. *Clim. Dyn.* **30** 887
[12] Behling H 2003 *Veget. Hist. Archaeobot.* **12** 263
[13] Vuille M, Bradley R S, Werner M, Healy R and Keimig F 2003 *J. Geophys. Res.* **108** doi:10.1029/2001JD002038
[14] Cruz F W C, Burns S J, Karmann I, Sharp W D, Vuille M, Cardoso A O, Ferrari J A, Dias P L S and Viana O 2005 *Nature* **434** 63
[15] Wang X, Auler A S, Edwards R L, Cheng H, Ito E, Wang Y, Kong X and Solheid M 2007 *Geophys. Res. Lett.* **34** doi:10.1029/2007GL031149
[16] Rasmussen S O et al. 2006 *J. Geophys. Res.* **111** doi:10.1029/2005JD006079
[17] Behling H 2002 *Palaeogeogr Palaeocl* **177** 19
[18] Behling H, Pillar V D, Orlóci L, Bauermann S G 2004 *Palaeogeogr Palaeocl* **203** 277
[19] Melo M L D and Marengo J A 2008 *Holocene* **18** 501
[20] McManus J F, Francois R, Gherardi J -M, Keigwin L D and Brown-Leger S 2004 *Nature* **428** 834
[21] Arz H W, Pätzold J, Wefer G 1999 *Earth. Planet. Sc. Lett.* **167** 105
[22] Stouffer R J et al. 2006 *J. Climate* **19** 1365
[23] Zhang R and Delworth T L 2005 *J. Clim.* **18** 1853
[24] Rohling E J and Pälike H 2005 *Nature* **434** 975
[25] Baker P A, Seltzer G O, Fritz S C, Dunbar R B, Grove M J, Tapia P M, Cross R B, Rowe H D and Broda J P 2001 *Science* **291** 640
[26] Araujo A G M, Neves W A, Piló L B and Atui J P V 2005 *Quaternary Res.* **64** 298
[27] Conkright M E, Locarnini R A, Garcia H E, O'Brien T D, Boyer T P, Stephens C and Antonov J I 2002 World ocean atlas 2001: Objective analyses, data statistics, and figures CD-ROM documentation, *Internal Rep. 17*, Natl. Oceanogr. Data Cent., Silver Spring MD
[28] Xie P and Arkin P A 1997 *Bull. Am. Meteorol. Soc.* **78** 2539
[29] Cruz F W C et al. 2009 *Nature Geosci.* **2** 210
[30] Berger A and Loutre M F 1991 *Quat. Sci. Rev.* **10** 297

IOP Conf. Series: Earth and Environmental Science **9** (2010) 012008 doi:10.1088/1755-1315/9/1/012008

Comparison of diatom records of the Heinrich Event 1 in the Western North Atlantic

Isabelle M Gil[1,2,3*], Lloyd D Keigwin[2] and Fatima G Abrantes[1]

[1]LNEG-Marine Geology Unit, Estrada da Portela – Alfragide, Apartado 7586 - 2720-866 Amadora, Portugal
[2]Woods Hole Oceanographic Institution, McLean 207A, MS#08, Woods Hole - Ma. 02543, USA
[3]CIMAR-Associate Laboratory, Rua dos Bragas, 289 - 4050-123 Porto, Portugal

E-mail: *igil@whoi.edu

Abstract: Heinrich event 1 (H1) is a climate event resulting from the release into the North Atlantic of a huge volume of sea ice and icebergs from the northern hemisphere ice sheets. We present here high-resolution diatom records from the Bermuda Rise (Sargasso Sea) and the Laurentian Fan (South of Newfoundland) to assess its impacts on North Atlantic surface circulation and its timing. The event is composed of three phases: the two first correspond to major pulses of Ice Rafted Debris (IRD) and the third phase relates to its immediate aftermath. At both sites, diatom abundances start to rise at 16.8 ka. This increase is marked by diatom species thriving in cold environments over the Laurentian Fan, while brackish and fresh water diatom species characterize this increase over the Bermuda Rise. This last record implies icebergs migration to subtropical latitudes and nutrient-rich meltwater to support such diatom productivity. During the second phase of the event, both sites record maximum diatom abundances, when sea ice conditions are dominant over the Laurentian Fan and the contribution of brackish-fresh water diatoms culminates over the Bermuda Rise. The persistence of lower salinity related diatom species over the Bermuda Rise suggests a continuous injection of cold fresh water by cold-core rings, as observed nowadays. The last phase witnesses the persistence of low salinity water over the Laurentian Fan, while a decrease towards the disappearance of diatoms in the sediment occurs over the Bermuda Rise. Regarding the relationship between IRD and diatom abundance, it appears that a critical amount of icebergs is necessary to stimulate diatom productivity. The disturbances induced by H1 appear to end ~14.6 ka over the Bermuda Rise, while over the Laurentian Fan, the high diatom production persists until 14.1 ka and the salinity anomaly until 13.8 ka.

1. Introduction

We propose to describe the environmental conditions of the North Atlantic from 19 to 14 ka. The main climatic event for that period is Heinrich Event 1 (H1). It is characterized by the release of icebergs and a consequent amount of meltwater into the North Atlantic [1]. Heinrich events are revealed by the presence of lithic grains, known as Ice Rafted Debris (IRD), in numerous sediment cores of the North Atlantic [2]. Because the sediments corresponding to this event are largely composed of IRD in the belt of maximum accumulation, it is difficult to get a detailed well-dated description of the event. Here, we present high-resolution diatom records from two western North Atlantic sites, the Bermuda Rise and the Laurentian Fan. At these sites, planktonic diatoms (microscopic algae) trace the conditions of surface waters (euphotic zone), in particular surface productivity, salinity and temperature changes. The comparison of diatoms at these two sites separated by 10° in latitude will also help to show the spatial extent of H1 across the northwest Atlantic.

2. Oceanographic setting

Both sites have high deposition rates (60.5 cm/ka for the Bermuda Rise and 44.1 cm/ka for the Laurentian Fan sites, respectively) that are attributed to the Gulf Stream system and deep recirculating gyres [3] over the Bermuda Rise and to the St Lawrence River sediment supply over the Laurentian Fan. The present day primary productivity at the Bermuda Rise site is low because of the low nutrient content of surface waters [4]. However, two mesoscale processes can enhance productivity: (a) mode water eddies, which activate upwelling, creating a sufficient nitrate flux to the photic zone to sustain new primary productivity [5]; (b) cold-core rings, transporting nutrient rich waters and biota from the subpolar gyre [6].

Over the Laurentian Fan, the nutrient-rich Labrador Current plays a major role in the productivity of the region [7]. There is also a strong temperature gradient, with the close juxtaposition of warm saline waters of subtropical origin and the flow of cold and less saline water of subpolar and arctic origin. Sea surface temperature (SST) controls the distribution of phytoplankton, with colder waters being more productive [8].

3. Material and data

Two gravity cores were studied (Fig. 1): OCE326GGC6 (GGC6 - 33°41.443'N; 57°34.559'W; 4541 m depth) and OCE326GGC14 (GGC14 – 43°03.959'N; 55°49.992'W; 3525 m depth).

The age model of GGC6 is based on its high correlation (magnetic susceptibility) with well-dated companion core OCE326GGC5 (GGC5; [9]). The correlation of the two records was made by using the Analyserie software [10] and the radiocarbon dates of GGC5 were transferred onto GGC6.

The age model of GGC14 is based on seven ^{14}C dates on foraminifera [11] and the age scale made using the Analyserie software [10].

All samples were radiocarbon dated at the National Ocean Sciences Accelerator Mass Spectrometer (NOSAMS) facility at Woods Hole Oceanographic Institution and converted to calendar years using CALIB4.3 calibration program [12]. The median of the error values is of 60 yrs for both cores. The sedimentation rates of GGC6 and GGC14 were not considered continuous.

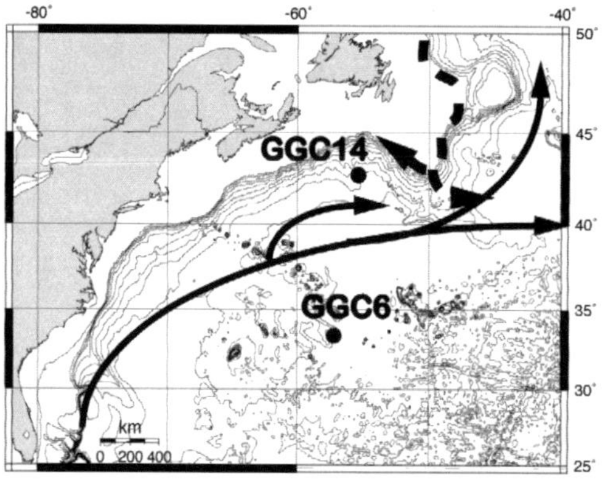

Figure. 1: Location map of studied cores GGC14 and GGC6. The plain arrows indicate the main pathway of the Gulf Stream and derived currents. The dashed arrows represent the Labrador Current.

In addition to diatom counts and assemblages, we present oxygen isotopes data from planktonic foraminifera [11] and Ice Rafted Debris (IRD) counts.

The diatom groups were constituted on the basis of the environmental conditions reported in studies from the North Atlantic (including the Nordic seas). Those studies associate diatom assemblages to the different currents of the North Atlantic and sea-ice conditions [13, 14, 15, 16].

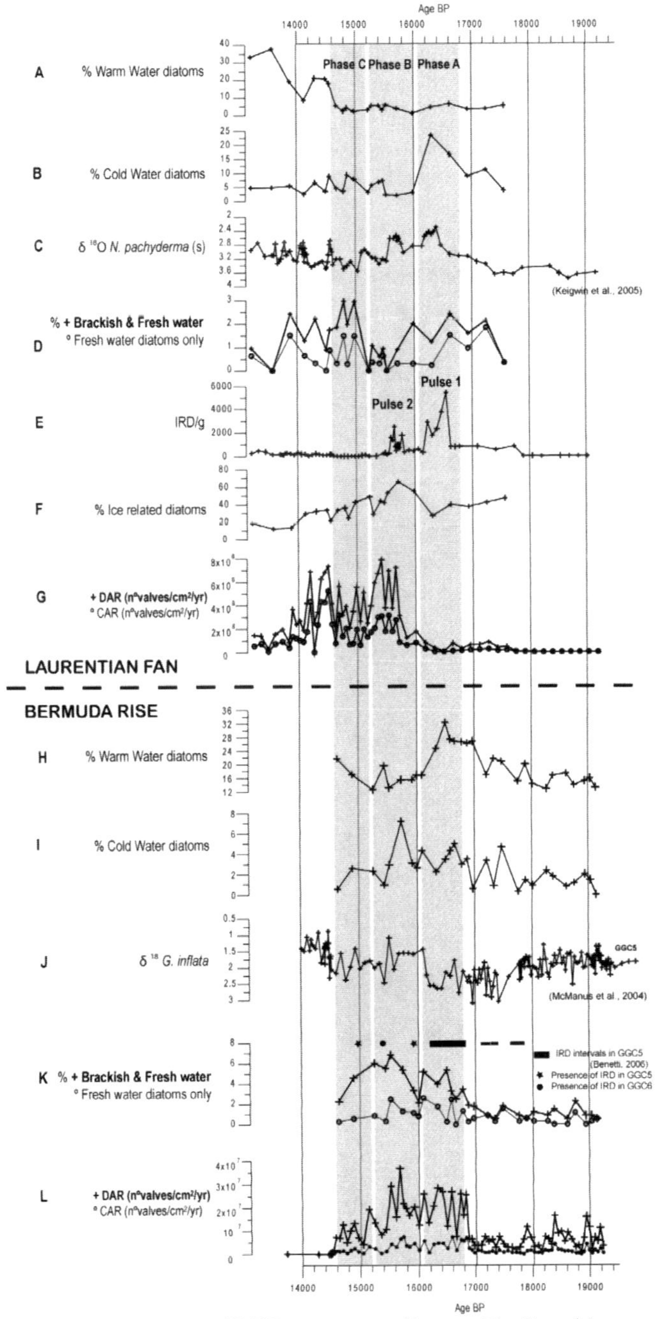

Figure. 2: Data from GGC14: A- % Warm water diatom, B- % cold water diatom, C- $\delta^{18}O$ (*N. pachyderma* (s)), D- % Brackish and fresh water diatoms (+) and % Fresh water diatom only (°), E- n° IRD/g, F- % ice related diatoms, G- Diatom Accumulation Rates (+) –DAR (n°valves/cm^2/kyr) and *Chaetoceros* Accumulation Rates (°) –CAR (n°valves/cm^2/kyr). Data from GGC6 and GGC5: H- % warm water diatom, I- % cold water diatom, J- $\delta^{18}O$ (*G. inflata*) in GGC5, K- % of brackish and fresh water diatoms (+) and % of fresh water diatom only (°) and intervals comprising IRD ([17] for GCC5), L- (+) DAR (n° valves/cm^2/kyr) and (°) CAR (n°valves/cm^2/kyr).

4. Results

Diatom Accumulation Rates (DAR) increase abruptly at 16.8 ka at the Bermuda Rise (Fig. 2 L) and remain high until 15.4 ka. In contrast, at the Laurentian Fan site they slowly start to increase at 16.5 ka (Fig. 2 G) until registering a period of high DAR from 16 to 14.1 ka (with a minimum from 15.8 to 15.2 ka). This increase in diatom production is sustained by the increase in cold, brackish and fresh water species (Fig. 2 I and K) at the Bermuda Rise, while at the Laurentian Fan, cold-water species dominate (Fig. 2 B).

Both sites register their maximum in DAR between 15.7 and 15.2 ka. This period is characterized by the maximum contribution of brackish and fresh water diatoms over the Bermuda Rise (Fig. 2 K), while at the Laurentian Fan this group registers a minima and the contribution of ice related diatom group peaks (Fig. 2 B and F).

From 15.2 to 14.6 ka, both sites register lower abundances, with at the Bermuda Rise a decline of brackish and fresh water diatoms (Fig. 2 K) and a rise in warm water diatoms (Fig. 2 A). Over the Laurentian Fan the minimum in total diatoms is characterized by a rise in brackish and fresh water diatoms and a decrease in ice related diatoms (Fig. 2 D and F).

At 14.6 ka, there are almost no diatoms preserved in the Bermuda Rise sediments and the contribution of warm water diatoms increases over the Laurentian Fan (Fig. 2 H).

IRD abundance increased in the sediments of both sites from ~17.8 ka. A first major input is simultaneously recorded at both sites from 16.8 to 16.2 ka (pulse 1) (Fig. 2 E and K). Between 16 and 15.4 ka, a second major input (pulse 2, marked by a minimum at ~15.7 ka) is recorded over the Laurentian Fan (Fig. 2 E), corresponding to two IRD levels at 15.4 ka (GGC5 and GGC6) and at 15.9 ka (GGC5) over the Bermuda Rise. An additional level in IRD in GGC5 is observed at 15 ka (Fig. 2 K).

Finally, surface water over the Laurentian Fan had lower $\delta^{18}O$ at ~16.3, 15.6, 15.1, 14.6 and 14.1 ka, with the first three more accentuated (Fig. 2 C). In contrast, the $\delta^{18}O$ record from the Bermuda Rise shows mostly maxima from 16.1 to 14.6 ka (Fig. 2 J).

5. Discussion

The diatom data obtained from the two studied sites indicates that H1 is composed of three phases (A, B and C; Fig. 2), with the first two being initiated by major IRD pulses and the last phase being marked by the persistence of a low salinity anomaly. However, it is important to note that IRD started to appear earlier in the sediments of both sites (from ~17.8 ka), when a notable rise in cold water diatoms over the Bermuda Rise was recorded (no diatom assemblage available for the Laurentian Fan site). It likely announced the following major inputs characterizing phases A and B.

Phase A occurred from ~16.5/16.8 to 16.1 ka. It is initiated by the IRD pulse 1 and marked by an abrupt increase in diatom production over the Bermuda Rise. Over the Laurentian Fan, the diatom production also tended to increase, but the abundance of diatom mainly contained in the fine fraction of the sediment may be diluted by the simultaneous occurrence of IRD. Comparing the two sites, a much more accentuated cooling is indicated over the Laurentian Fan. The notable increase in cold-water diatom species over the Bermuda Rise (from 0.2% to 4.6%) was probably related to the presence of ice, while the enhanced contribution of warm water diatoms likely reflected the increase in the local diatom production expected at this latitude.

The coincidence of the arrival of icebergs (evidenced by the IRD) and the increase in diatom production suggests that icebergs stimulated diatom productivity [18]. Although there is no modern analog in the North Atlantic to sustain this hypothesis, several examples of primary productivity generated by the presence of icebergs in the Southern Ocean (SO) have been reported [19, 20].

Phase B, from ~16.1 to 15.2 ka, corresponds to the maximum in diatom production at both sites. The second IRD pulse (pulse 2) was less important than pulse 1 and the presence of IRD in Bermuda Rise sediments is sparse. Over the Laurentian Fan, there were fewer low salinity indicators, a peak in abundance of ice related diatoms, and a marked decrease in $\delta^{18}O$ suggesting again the presence of ice over the site. Over the Bermuda Rise, the diatom production was clearly linked to an increase in low salinity water. Brackish and fresh water diatoms grew in situ or were transported by icebergs. Nowadays, such an increase in diatom productivity is generated by cold-core rings and their input in nutrients [6]. The melting of icebergs must have amplified these oceanographic processes. Interestingly, the period between the two IRD pulses over the Laurentian Fan corresponds to a lower abundance of brackish and fresh water diatoms at ~16 ka at the Bermuda Rise, establishing a link between the sites and confirming the origin of the low salinity water.

Finally, phase C took place between 15.2 and 14.6 ka. Over the Bermuda Rise, the decrease in brackish and fresh water diatoms was linked to the progressive disappearance of diatoms in the sediments and to a warming of the surface water starting at 15.2 ka. Such warming may mark the origin of the Bølling warming or just reflect a return to ambient conditions expected for that latitude. Although the salinity anomaly was diminished, it could still have contributed to the local production of diatoms. Another possibility is that diatom productivity was also amplified by the silica leakage from the Southern Ocean (SO) during deglaciation [21]. Silica rich water such as the Antarctic Intermediate Water (AAIW) could have been brought to the euphotic zone through mode-water eddies [18]. In contrast, the Laurentian Fan site was still under the influence of melt water and the transition to warmer conditions started later at 14.6 ka (also observed at 14.6 ka in the GRIP ice core; [22]). From 14.6 to 14.1 ka, a rise in DAR (linked principally to warm water diatoms and to brackish/fresh diatoms) was recorded over the Laurentian Fan. As H events are defined by the presence of IRD in the sediment, it is difficult to relate directly this increase to the consequences of H1. However, the higher DAR values could have resulted from the melting of clean icebergs (not containing IRD) and/or sea-ice, due to the warming of the water. It is also possible that a large part of the ice from the Gulf of St Lawrence melted and the water reached the Laurentian Fan in important quantities. The icebergs and/or the sea-ice would have only prevailed at northern latitudes by this time and it would explain why those conditions generating diatom productivity did not affect the Bermuda Rise. Besides, as the high diatom production persisted until 14.1 ka and the lower salinity indicators were present until 13.8 ka, this could also testify in favor of a restricted contribution of the Laurentian ice sheet to the melt water pulse 1a [23].

The icebergs and their meltwater likely stimulated an immediate increase in productivity over the Bermuda Rise, while over the Laurentian Fan the increase in diatom abundance was progressive during pulse 1 (major pulse recorded at both sites) but fast during pulse 2. These suggest that to stimulate the productivity, there is a need for a critical amount of icebergs, meltwater and associated turbulence. However, the release of finer particles with the IRD can also cause unclear water and prevent the development of diatoms.

6. Conclusion

The simultaneity of the successive environmental changes recorded at two high-resolution sites on either side of the Gulf Stream confirms that H1 is composed of a sequence of events. Three main phases are distinguished, with the first two defined by IRD pulses and the last one corresponding to the impacts of those IRD pulses. The phasing of these events is particularly evidenced around 16 ka, between the two IRD pulses, when the low salinity anomaly from the Bermuda Rise is clearly diminished.

The perturbation induced by H1 ends at ~14.6 ka over the Bermuda Rise and at 14.1 ka over the Laurentian Fan.

Acknowledgements

Support for this research was provided the Fundação para a Ciência e Tecnologia-FCT through the grant BPD/27214/2006 to I. M. Gil.

Financial support to attend this conference ca e from PAGES and CIIMAR.

References

[1] W. Broecker, G. Bond, M. Klas, E. Clark, and J. McManus, Origin of the northern Atlantic's Heinrich events. Climate Dynamics 6 (1992) 265-273.

[2] S.R. Hemming, Heinrich events: Massive late pleistocene detritus layers of the North Atlantic and their global climate imprint. Reviews of Geophysics 42 (2004) -.

[3] E.P. Laine, Hollister, C. D., Geological effects of the Gulf Stream system on the northern Bermuda Rise. Marine Geology 39 (1981) 277-310.

[4] G.R. Heath, Dissolved silica and deep-sea sediments. in: W.W. Hay, (Ed.), Studies in paleo-oceanography, 1974, pp. 77-93.

[5] D.J. McGillicuddy, L.A. Anderson, N.R. Bates, T. Bibby, K.O. Buesseler, C.A. Carlson, C.S. Davis, C. Ewart, P.G. Falkowski, S.A. Goldthwait, D.A. Hansell, W.J. Jenkins, R. Johnson, V.K. Kosnyrev, J.R. Ledwell, Q.P. Li, D.A. Siegel, and D.K. Steinberg, Eddy/wind interactions stimulate extraordinary mid-ocean plankton blooms. Science 316 (2007) 1021-1026.

[6] The Ring Group, Gulf Stream cold-core rings: their physics, chemistry, and biology. Science 212 (1981) 1091-1100.

[7] S. Dutkiewicz, Follows, M., Marshall, J., Gregg, W. W., Interannual variability of phytoplankton abundances in the North Atlantic. Deep-Sea Research II 48 (2001) 2323-2344.

[8] J.U. Umoh, Seasonal and interannual variability of sea temperature and surface heat fluxes in the Northwest Atlantic, Dalhousie University, Halifax, 1992, pp. 237.

[9] J.F. McManus, François, R., Gherardi, J.-M., Keigwin, L. D., Brown-Leger, S., Collapse and rapide resumption of Atlantic meridional circulation linked to deglacial climate changes. Nature 428 (2004) 834-837.

[10] D. Paillard, L. Labeyrie, Yiou, P., Macintosh program performs time-series analysis. Eos Trans. 77 (1996).

[11] L.D. Keigwin, J.P. Sachs, Y. Rosenthal, and E.A. Boyle, The 8200 year BP event in the slope water system, western subpolar North Atlantic. Paleoceanography 20 (2005) -.

[12] M. Stuiver, Reimer, P. J., Bard, E., Beck, W., Burr, G. S., Hughen, K. A., Kromer, B., McCormac, G., van der Plicht, J., Spurk, M., INTCAL98 radiocarbon age calibration, 24,000-0 cal BP. Radiocarbon 40 (1998) 1041-1083.

[13] N.G. Maynard, Relationship between diatoms surface sediments of the Atlantic Ocean and the biological response and physical oceanography of overlying waters. *Paleobiology* 2 (1976) 99-121.

[14] N.K. Karpuz, Schrader, H., Surface sediment diatom distribution and Holocene paleotemperature variations in the Greenland, Iceland and Norwegian Sea. Paleoceanography 5 (4) (1990) 557-580.

[15] G.R. Hasle, Syversten, E. E., Steidinger, K. A., Tangen, K., Marine diatoms. in: C.R. Tomas, (Ed.), Identifying marine diatoms and dinoflagellates, Academic Press, Inc., San Diego, 1996.

[16] A. Justwan, and N. Koc, A diatom based transfer function for reconstructing sea ice concentrations in the North Atlantic. Marine Micropaleontology 66 (2008) 264-278.

[17] S. Benetti, Late Quaternary sedimentery processes along the western North Atlantic margin, School of Ocean and Earth Sciences, University of Southampton, Southampton, 2006, pp. 188.

[18] I.M. Gil, L.D. Keigwin, and F.G. Abrantes, Deglacial diatom productivity and surface ocean properties over the Bermuda Rise, northeast Sargasso Sea. Paleoceanography 24 (2009).

[19] K.L. Smith, B.H. Robison, J.J. Helly, R.S. Kaufmann, H.A. Ruhl, T.J. Shaw, B.S. Twining, and M. Vernet, Free-drifting icebergs: Hot spots of chemical and biological enrichment in the Weddell Sea. Science 317 (2007) 478-482.

[20] J.N. Schwarz, and M.P. Schodlok, Impact of drifting icebergs on surface phytoplankton biomass in the Southern Ocean: Ocean colour remote sensing and in situ iceberg tracking. Deep-Sea Research Part I-Oceanographic Research Papers 56 (2009) 1727-1741.

[21] L.I. Bradtmiller, R.F. Anderson, M.Q. Fleisher, and L.H. Burckle, Opal burial in the equatorial Atlantic Ocean over the last 30 ka: Implications for glacial-interglacial changes in the ocean silicon budget. Paleoceanography 22 (2007).

[22] S.O. Rasmussen, K.K. Andersen, A.M. Svensson, J.P. Steffensen, B.M. Vinther, H.B. Clausen, M.L. Siggaard-Andersen, S.J. Johnsen, L.B. Larsen, D. Dahl-Jensen, M. Bigler, R. Rothlisberger, H. Fischer, K. Goto-Azuma, M.E. Hansson, and U. Ruth, A new Greenland ice core chronology for the last glacial termination. Journal of Geophysical Research-Atmospheres 111 (2006) -.

[23] E. Bard, B. Hamelin, M. Arnold, L. Montaggioni, G. Cabioch, G. Faure, and F. Rougerie, Deglacial sea-level record from Tahiti corals and the timing of global meltwater discharge. Nature 382 (1996) 241-244.

IOP Conf. Series: Earth and Environmental Science 9 (2010) 012009 doi:10.1088/1755-1315/9/1/012009

Spatial patterns in Central Asian climate and equilibrium line altitudes

S Rupper[1], M Koppes[2]

[1] Brigham Young University, Dept. of Geological Sciences, Provo, Utah, U.S.A
[2] University of British Columbia, Dept. of Geography, British Columbia, Canada

summer_rupper@byu.edu

Abstract. A suite of general circulation model (GCM) simulations and a glacier equilibrium line altitude (ELA) model are compared to reconstructed glacier advances from geomorphic data and used to test the sensitivity of Central Asian glaciers to simulated climate changes at the Last Glacial Maximum (LGM). Results highlight temperature changes as being the most important influence on glacier ELA changes during the LGM. With the exception of the southern Himalaya, for much of Central Asia there is consistency between GCMs for simulated LGM temperature changes, with a mean cooling of 4° C. Further research will necessarily need to focus on detailed analysis of the inter-model differences in temperature in the southern Himalaya, and acquiring additional paleoclimate proxies in the region in order to further constrain the GCMs.

1. Motivation

Central Asia has a special role in unraveling the relation between climate and glacier response, in part because the apparent simplicity of its continental climate - dominated by the westerly storm track, the Siberian High in the winter, and the South Asian Monsoon in the summer [1] - has made regional deviations more obvious. However, the sheer size of the region has made it difficult to define the glacial and climate histories of Central Asia as a whole. It is important to focus fieldwork in those regions that have the greatest potential for defining the regional climate and glacier histories.

General circulation model (GCM) simulations can be used to guide field efforts in Central Asia, to test the sensitivity of glacier mass balance to plausible changes in climate, and as input in glacier models for comparison to geomorphic data. We test the sensitivity of Central Asian glaciers to changes in both precipitation and ablation using the Paleoclimate Model Intercomparison Project [2] suite of GCM simulations and a glacier equilibrium line altitude (ELA) model. The ELA model utilizes the traditional approach to estimating mass balance by assuming accumulation is the annual average of all precipitation above the zero degree isotherm and ablation is proportional to temperature (i.e., ablation is a function of positive degree days). The results of the GCM simulations and ELA model are compared to the geomorphic and chronologic data of glacial advances during the global Last Glacial Maximum (LGM, heretofore defined as approx. 21 kyr B.P.) in Central Asia to assess model-data discrepancies and inter-model variability.

The two main goals of this preliminary study are to 1) reconcile the record of ELA changes with the pattern of climate change around the region and 2) test the sensitivity of ELAs to changes in LGM boundary conditions.

IOP Conf. Series: Earth and Environmental Science **9** (2010) 012009 doi:10.1088/1755-1315/9/1/012009

2. Variable history of the Last Glacial Maximum across Central Asia

A growing body of work in the past two decades has taken advantage of recent advances in geochronometric techniques, especially the widespread use of cosmogenic radionuclides for dating moraines. Although sparse in coverage, and acknowledging uncertainties in the dating techniques used, the ~20 chronologic studies conducted to date suggest that glacier response to climate changes during the global LGM was highly variable across Central Asia. For ease of analysis, we subdivide the geography of Central Asia into three major zones, western, northern and eastern, each demonstrating significantly different climate patterns today and in the past [3,4].

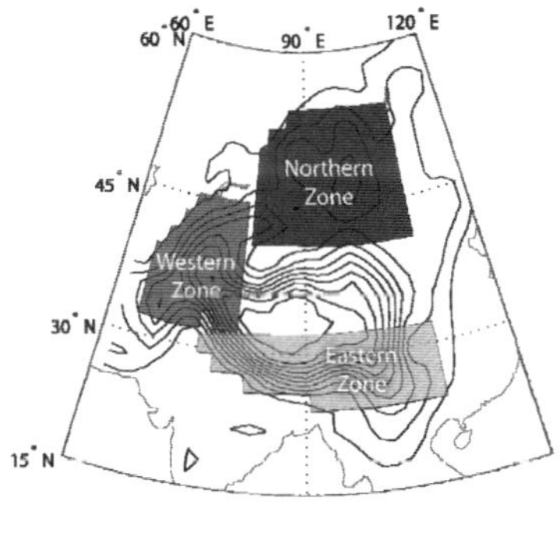

Figure 1: Subdivision of Central Asia into 3 zones used in this study. Figure modified after [3]. Contour interval is 500 m.

The western zone includes the Kyrgyz Tien Shan south to the Karakoram and east to central Tibet. Evidence of large glacial advances during the LGM is lacking in much of the region, with most of the LGM advances restricted to within a few km of extant glaciers, and often nested within much larger earlier advances. For example, in the Pamir Range, moraines from smaller LGM advances are bracketed within larger pre-LGM moraines [5], corresponding to an ELA depression of 280-320 m during the LGM. In the Hunza valley and at Nanga Parbat, glaciers advanced less than 10 km from their current termini during the LGM [6,7]. LGM advances in the Kyrgyz Tien Shan were restricted to the cirques, close to the modern glaciers [8]; here, the ELA depression during the LGM was at most 150 m. In the one exception to this trend, dates are reported for two large LGM advances in the Hindu Kush [9], which correspond to an ELA depression of approximately 1000 m.

The northern zone is defined by the area from central Mongolia to China's Xinjiang Tien Shan. Here there is evidence of substantial glacial advances during the LGM as well as pre-LGM. For example, large advances occurred around 19-17 ka in the Darhad Basin of northern Mongolia [10], in the Die Han Je Le Gou drainage in the Xinjiang Tien Shan [11], and in the Russian and Mongol Altai [12]. The ELA lowered during these advances by 900-1200 m. Throughout the northern region, changes in ELA appear to increase with latitude, from -500 m in the Xinjiang Tien Shan to -1000-1200 m north of 51° N [12,13]. The difference between the maximum ELA depression (presumed to be pre-LGM) and that of the LGM is only 25-100 m.

The glacial history of the eastern zone, which comprises the southern Himalaya and eastern Tibet, appears to be more variable than in the other two zones. Several studies suggest that the glacial advances in the LGM were significantly smaller than those during pre-LGM times, particularly in mountains of the SE Tibetan Plateau [14] and western China [15], and around Mt. Everest [16]. Other studies found no evidence at all of any significant glacial advance during the LGM in eastern Tibet [14,17]. However, others suggest that the largest advance at Kanchenjunga, in the eastern Himalaya, occurred during the LGM [18]. Thus, the exact timing of extensive glacial advance in this zone is still under debate. The corresponding maximum ELA depressions in the region increased from west to east as well as from north to south, ranging from 300-500 m on the Tibetan plateau to >1000 m in the eastern Himalaya [12,13].

In summary, a pattern of large spatial variability in LGM glacier advance across Central Asia is emerging from the geomorphic data, with smaller glacier advances in the western zone, substantial advances in the northern zone, and inconsistencies in the eastern zone. Since glaciers advance and retreat in response to changes in climate, the spatial variability in glacier ELAs and length changes at the peak of the LGM (21 ka) suggest spatial variability in climate also occurred.

3. Simulated Last Glacial Maximum climate

We focus here on analyzing the precipitation and temperature output from the 12 highest resolution PMIP GCMs because of the important link between these two variables and glaciers [2]. Table 1 lists the mean changes in temperature and precipitation averaged from all GCMs over each of the three zones. Any differences between the model outputs is highlighted by using the standard deviation divided by the absolute value of the mean (of either summertime (JJA) temperature or annual precipitation) for all model simulations, at each grid point. These values represent an index of how consistent the model outputs are, i.e., an index of variability (IV in Table 1). An index of variability of <1 indicates regions where the standard deviation across the models is small relative to the mean change in summertime temperature or precipitation and suggests consistency in the model simulations. Despite the large variability between the GCM model simulations (discussed below) these models can nonetheless be used to evaluate the range of predicted climate response to LGM boundary conditions.

3.1. Patterns in temperature change

All the GCMs simulate a decrease in summertime (JJA) temperatures at 21 ka in the northern and western zones. In contrast to this general decrease in temperatures, the temperature changes are more variable along the southern Himalaya, including in the eastern zone, ranging between +3 °C and -5 °C. Figure 2a shows the mean temperature change from the 12 highest resolution GCMs (see also Table 1 for averages over each zone). The output for all 12 models was linearly interpolated to the same grid, and then averaged at each grid point. The mean of the models shows a decrease in temperature everywhere with the magnitude of the temperature change generally increasing with increasing latitude and over the Tibetan Plateau. The spread in the model results, i.e., the index of variability, is illustrated in Figure 2c. For most of Central Asia the index of variability for modeled temperatures is less than 1 (blue color in Figure 2c), indicating a fairly robust response of the simulated temperatures to changes in boundary conditions at 21 ka. In only a small portion of the eastern Himalayas is the temperature change more variable (red areas in Figure 2c). As will be shown below, such patterns in temperature change are important for explaining the modeled ELA changes across Central Asia.

Table 1: Changes in climate variables and ELAs averaged for all GCM simulations for the eastern, western, and northern zones between the present and 21 ka. T is the mean summer (JJA) air temperature, P is mean annual precipitation, ELA is the equilibrium-line altitude, IV is the index of variability (absolute values of the standard deviation divided by the mean change).

	ΔT (°C)	**IV for T**	**ΔP (m yr^{-1})**	**IV for P**	**ΔELA (m)**	**ΔELA for ΔT only (m)**
Eastern Zone	-3	1	-0.1	>2	-500	-440
Western Zone	-4	0.5	-0.01	>2	-1000	-900
Northern Zone	-4	0.6	-0.08	0.25	-1000	-910

3.2. Patterns in precipitation change

There is a large spread in the GCM simulated changes in annual precipitation for much of Central Asia, with the exception of the northern zone. In the northern zone, annual precipitation decreases in all GCMs averaged between 1.5 and 20 cm/yr (index of variability < 1). The tendency of all the models towards an overall decrease in precipitation is likely due to the simultaneous decrease in temperature, as given by the Clausius-Clayperon relation (i.e., for constant relative humidity, moisture content increases with temperature [19]). On the other hand, precipitation changes at 21 ka in the western and eastern zones are highly variable from model to model (index of variability > 2, Figure 2d), and even the sign of the change is not consistent (Table 1, Figure 2b). Such a large degree of variability between the model outputs may be due in part to differences in how each of the GCMs is simulating changes in the Mediterranean storm track (particularly in the western zone) and in the South Asian monsoon (in the eastern zone) in response to LGM boundary conditions.

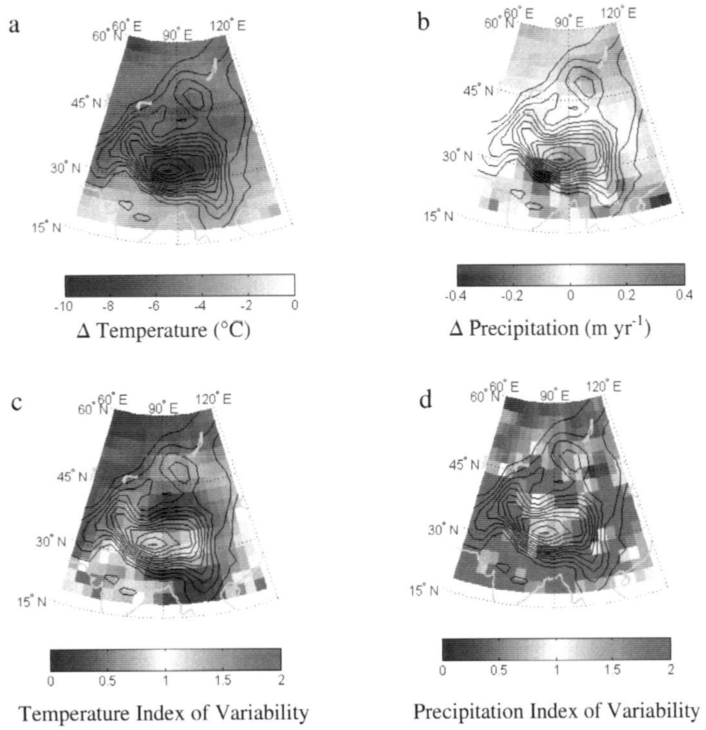

Figure 2: (a) mean change in summertime (JJA) temperature, (b) mean change in precipitation, (c) index of variability in summertime temperature, and (d) index of variability in precipitation for 12 PMIP GCM simulations.

In summary, the simulated temperature changes are consistent for the northern and western zones, an indication that summertime temperatures are a robust response of the climate to changes in LGM boundary conditions in those regions. Temperature changes in the southern Himalaya are more variable, however, suggesting that there is a high degree of uncertainty in the response of the climate to LGM boundary conditions in regions dominated by the South Asian monsoon. There is an even larger spread in precipitation changes simulated by the GCMs across much of Central Asia, with the exception of the northern zone. Thus, precipitation changes are far more uncertain than temperature changes across the region.

4. Modeled spatial pattern in ELA change

To understand the impact of the simulated climate changes on glacier mass balance, and to compare to the paleoclimate data, the GCM output must be converted into changes in ELA. To do this we use GCM temperature and precipitation as input for an ELA perturbation model [20], modified to use a positive degree day (PDD) model for ablation [21], as described in [4]. The exact equation used to solve for the change in ELA for a step change in climate is as follows:

$$\Delta A_c = \Delta A_b + \frac{\partial A_b}{\partial z} \Delta ELA, \qquad \text{Equation 1}$$

where Δ refers to the change from a reference climate, A_c is the annual average accumulation, A_b is the annual average ablation, $\frac{\partial A_b}{\partial z}$ is the gradient in ablation, and ELA is the equilibrium line altitude. Ablation is assumed to be proportional to positive degree days (the sum of air temperatures greater than zero). Details on the derivation of this equation can be found in [4].

Average ELA changes for all GCMs in the three separate zones are listed in Table 1. In the northern zone, the model results corresponded quite well to the paleoclimate record of ELA depression during the LGM, which averaged 900-1200 m. [13]. By comparison, the paleo-ELA depressions recorded during the LGM in the western zone were much smaller than the model results, with the exception of the Hindu Kush [7], where the ΔELA was similar to the model outputs, ~1000m. Assuming that the paleo-ELAs reconstructed from moraine records for the western zone are reliable, the GCMs are significantly overestimating the ELA decrease in the western region. In the eastern zone, the relationship between mapped and modeled ELA changes is harder to assess, in part due to the high degree of variability between the models (see below), but also due to disagreement in the geomorphic data about the extent of LGM glaciation in the region [14,18].

There are important discrepancies between the patterns of ELA changes for each of the different GCMs, as can be seen in Figure 3. The largest discrepancy between models occurs in the southern Himalayas. Even the sign of the ELA change is not consistent between models in this region, ranging between -1000 m and +500 m. This coincides with the same region where the sign of the temperature changes was also highly variable between the models, and where the paleoclimate data is most inconsistent. All models predict a lowering of ELAs in the northern and western zones, but the pattern of this lowering is different for different GCMs. For example, the ΔELAs for the GFDL GCM generally increase with latitude (Figure 3a). By comparison, the ΔELAs for the LMCDEL-5 GCM shows the largest change over the interior of the Tibetan Plateau and decreasing changes at higher latitudes (Figure 3b). Therefore, while the sign of the ELA changes are consistent, the pattern of ELA changes is different between the models.

While there are obvious uncertainties in the GCM precipitation and temperature output, both in magnitude and spatial pattern, the large spread in output provides a good test of ELA sensitivity to temperature versus precipitation. In particular, the pattern in ELA changes appears to reflect the pattern in temperature changes. To determine to what degree patterns in temperature change are controlling the pattern in ELA model output, the change in ELA is calculated for change in temperature only (i.e., no change in precipitation). If only the temperature is modified, the changes in ELA for all three zones are approximately 80-90% of the total change in ELA. In other words, despite differences in the magnitude of the ELA response, ELAs across Central Asia using our ELA model are most sensitive to simulated changes in temperature, which controls ablation, and much less to changes in precipitation, which controls accumulation.

The strong relationship between temperature and ELA patterns is not surprising given that the simple PDD method assumes that ablation is directly related to temperature. Physically, of course, it is a net input of heat that is actually responsible for the melting or sublimation of snow and ice. To address ablation more realistically, we also compared our PDD results to a full surface energy- and mass-balance model (SEMB model), as presented in [3]. Complete description and analysis of the results of the SEMB model are beyond the scope of this paper, but a brief comparison does provide a check on the results derived from the simpler PDD model. First, the overall pattern in ELA changes using the SEMB approach is very similar to that of the PDD approach for each of the GCM simulations. Where the two methods disagree is in regions where sublimation dominates ablation [3]. Second, we also modeled the ELA changes for a change in ablation only using the SEMB approach. If precipitation does not change, the change in ELAs in all melt-dominated regions account for more than 80% of the total change in ELA. Both of these results, using the SEMB model, are consistent with the results from the PDD model presented above. In summary, where melt dominates ablation, the PDD approach to estimating ablation can adequately capture the patterns of ELA change seen using a full surface energy- and mass-balance approach.

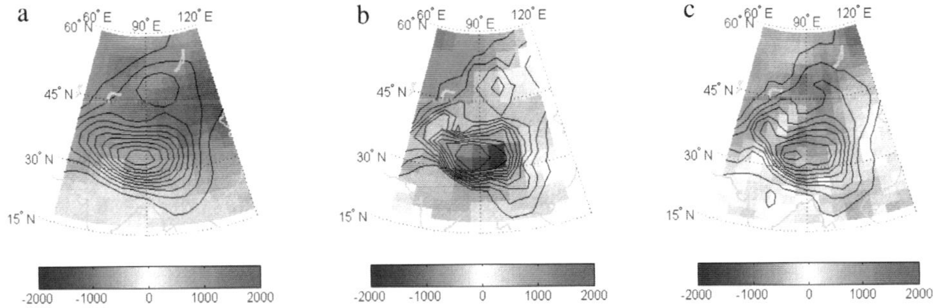

Figure 3: Changes in ELAs for present-day minus 21 ka for three GCMs that illustrate the spread in the results, calculated using the PDD approach to ablation: a) GFDL, b) LMCELMD-5, c) BMRC-2.

5. Discussion and conclusion

Since the pattern in temperature changes appear to play defining role in the modeled ELA changes, insight is needed to understand why these patterns in temperature change occur. In particular, the connection between the changes in radiation and the temperature response at the surface need to be prodded further. Preliminary analysis of the GCM outputs suggest that the decrease in temperature in the northern and western zones is due to the decrease in the longwave radiation flux, consistent with the idea that continental interiors are slave to changes in radiative forcing. This radiative cooling is due, at least in part, to a decrease in greenhouse gas concentrations and cloud fraction. Any differences in cloud fraction between the GCMs will influence the shortwave and longwave radiation fluxes, giving rise to variability in temperature changes. Therefore, while a decrease in greenhouse gases during the LGM would tend to cool the climate, variability in cloudiness between different GCMs can either amplify or dampen this cooling.

Further study is also needed to better understand the significant differences between the modeled and reconstructed ELAs. In particular, the modeled ELA changes in the western zone significantly overestimate the observed ELA depression during the LGM. Assuming the paleo-ELAs in this region are reliable, this discrepancy between models and data may suggest that sublimation dominated ablation in the west during the LGM. This would greatly increase ELA sensitivity to even small changes in precipitation. Any increase in precipitation sensitivity could compensate for the large change in ablation due to radiative changes. Further study using a full surface energy- and mass-balance approach will be required to test this hypothesis and determine whether the modeled ELAs overestimate the observed ELA depression during the LGM because of uncertainties in the ELA model, in the GCM temperature output, in the paleoclimate record, or in some combination of all three.

Results from [3] and [4] suggest that changes in temperature and precipitation in the eastern zone and the southern Himalayas are due to the dynamic response of the monsoon to radiative changes in the continental interior. The variability in the temperature changes and causes of those changes between models suggests that the response of the monsoon to LGM boundary conditions is highly variable between GCMs. In modern climate, monsoon intensity and duration is influenced by numerous factors, including land-sea temperature contrast, snow cover, and ENSO [22-27]. The monsoon response in the GCMs at 21 ka therefore likely depends on the exact response of each of these factors to LGM boundary conditions.

Understanding the exact reasons for the variability between the GCMs, as well as between the models and the paleoclimate data, will necessarily require diagnosing the model changes and sensitivity to those factors that influence the intensity and duration of the monsoon. In turn, a better understanding of past changes in the monsoon will help elucidate the relationship of the monsoon to glacier mass balance and glacier health, with important implications for the impact of future climate changes in the region. It is therefore important to couple the climate modeling efforts with additional glacial geomorphic mapping and age constraints in those regions directly influenced by the monsoon and the westerly storm tracks, particularly in the Himalayas and the mountains of the western zone, in order to better reconstruct both the glacial and climate histories in that region.

References

[1] Benn DI and Owen LA 1998 *Journal of the Geological Society of London* **155** 353-363
[2] Bonfils CB, Lewden D, and Taylor K 1998 *Summary documentation of the PMIP models* PCMDI Report
[3] Rupper SB and Roe GH 2008 *Journal of Climate* **21** 5384-5401
[4] Rupper SB, Roe GH, and Gillespe A 2009 *Quaternary Research* **72** 337-346
[5] Abramowski U, Bergau A, Seebacha D, Zech R, Glaser B, Sosin P, Kubik P W, and Zech W 2006 *Quaternary Science Reviews* **25** 1080–1096
[6] Phillips WM, Sloan VF, Shroder JF, Sharma P, Clarke ML, and Rendell HM 2000 *Geology* **28** 431-434
[7] Owen LA, Finkel RC, Caffee MW, and Gualtieri L 2002 *Geological Society of America Bulletin* **114** 593-604
[8] Koppes M, Gillespie A, Burke R, Thompson S, and Stone J, 2008 *Quaternary Science Reviews* **27** 846-866
[9] Owen LA, Kamp U, Spencer JQ, and Haserodt K 2002 *Quaternary International* **97–98** 41–56
[10] Gillespie A, Burke R, Komatsu G, and Bayasgalan A 2008 Quaternary Research **69** 169-187
[11] Clark D, Gillespie A, Bierman P, and Caffee M, 2001 *Geological Society of America Abastracts* **33** 441
[12] Lehmkuhl F, and Owen LA 2005 *Boreas* **34** 87–100
[13] Shi Y 2002 *Quaternary International* **97-98** 79-91
[14] Owen LA, Finkel RC, Barnard PL, Ma H, Asahi K, Caffee MW, and Derbyshire E 2005 *Quaternary Science Reviews* **24** 1391-1411
[15] Zhang W, Cui Z, and Li Y 2006 *Quaternary International* **154-155** 32-43
[16] Finkel RC, Owen LA, Barnard PL, and Caffee MW, 2003 *Geology* **31** 561–564
[17] Schäfer J M, Tschudi S, Zhizhong Z, Xihao W, Ivy-Ochs S, Wieler R, Baur H, Kubik P W, and Schlüchter C 2002 *Earth and Planetary Science Letters* **194** 287–297
[18] Tsukamoto S, Asahi K, Watanabe T, and Rink WJ 2002 *Quaternary International* **97-8** 57-67
[19] Wallace and Hobbs 2005 *Atmospheric Science An Introductory Survey* Academic Press
[20] Ambach W and Kuhn M 1985 *Z Gletscherkd Glazialgeol* **21** 311-317
[21] Braithwaite RJ 1995 *Journal of Glaciology* **41** 153-160
[22] Barnett T, Dumenil L, Schlese U, and Rockner E 1988 *Science* **239** 504-507
[23] Cohen J and Rind D 1990 *Journal of Climate* **4** 689-706
[24] Wang B, Wu R, and Fu X 2000 *Journal of Climate* **13** 1517-1536
[25] Gong G, Entekhabi D, and Cohen J 2002 *Journal of Climate* **16** 3917-3932
[26] Barlow M, Cullen H, and Lyon B 2002 *Journal of Climate* **15** 697-700
[27] Tippett M, Barlow M, and Lyon B 2003 *International Journal of Climatology* **23** 1421-1433

IOP Conf. Series: Earth and Environmental Science **9** (2010) 012010 doi:10.1088/1755-1315/9/1/012010

Understanding ENSO dynamics through the exploration of past climates

Steven J Phipps[1] **and Jaclyn N Brown**[2]

[1] Climate Change Research Centre, Faculty of Science, University of New South Wales, UNSW Sydney, NSW 2052, Australia
[2] CSIRO Marine and Atmospheric Research, GPO Box 1538, Hobart, TAS 7001, Australia

E-mail: s.phipps@unsw.edu.au

Abstract. The palaeoclimate record shows that significant changes in ENSO characteristics took place during the Holocene. Exploring these changes, using both data and models, provides a means of understanding ENSO dynamics. Previous modelling studies have suggested a mechanism whereby changes in the Earth's orbital geometry explain the strengthening of ENSO over the Holocene. Decreasing summer insolation over the Asian landmass resulted in a weakening of the Asian monsoon system. This led to a weakening of the easterly trade winds in the western Pacific, creating conditions more favourable for El Niño development. To explore this hypothesised forcing mechanism, we use a climate system model to conduct a suite of simulations of the climate of the past 8,000 years. In the early Holocene, we find that the Asian summer monsoon system is intensified, resulting in an amplification of the easterly trade winds in the western Pacific. The stronger trade winds represent a barrier to the eastward propagation of westerly wind bursts, therefore inhibiting the onset of El Niño events. The fundamental behaviour of ENSO remains unchanged, with the major change over the Holocene being the influence of the background state of the Pacific on the susceptibility of the ocean to the initiation of El Niño events.

1. Introduction

El Niño-Southern Oscillation (ENSO) is the dominant mode of internal variability within the coupled atmosphere-ocean system, with an irregular period of around two to seven years [1].

The background state of the tropical Pacific Ocean is characterised by a zonal sea surface temperature gradient, with a "warm pool" in the west. These warm surface waters give rise to atmospheric convection, driving easterly trade winds at the surface. These winds advect warm surface waters to the west and cause deeper, colder waters to upwell in the east, therefore maintaining the surface temperature gradient. This gives rise to a background state which is self-reinforcing, but which represents a state of unstable equilibrium. An El Niño event arises when this background state is perturbed. If the easterly trade winds are weakened in some manner, then the warm surface waters in the west flow eastwards. This leads to enhanced sea surface temperatures in the central and eastern Pacific, accompanied by strengthened atmospheric convection and enhanced precipitation.

The palaeoclimate record shows that significant changes in El Niño characteristics have taken place during the past 10,000 years [2–5]. Prior to around 7 ka[1], there was only weak, decadal-

[1] ka = thousand years before present

scale variability. "Modern" El Niño activity began around 5–7 ka, with a gradual strengthening of ENSO thereafter. There is evidence of a peak in intensity around 1-2 ka, although strong millennial-scale variability is also apparent [4].

These changes in El Niño have taken place during a relatively quiescent period in the Earth's climatic history, with no significant glaciation or deglaciation. Between 8 ka and the beginning of the pre-industrial era, atmospheric greenhouse gas concentrations were relatively constant, with the CO_2 concentration only varying between 260 and 280 ppm [6]. However, pseudo-cyclic changes in the Earth's orbital geometry have given rise to significant changes in the seasonal cycle of insolation. In particular, at northern mid-latitudes, the amount of insolation during the summer months has decreased by 30–40 Wm^{-2} between 8 ka and the present day.

Modelling studies have shown that the behaviour of ENSO is influenced by the seasonal cycle of insolation in the tropics, and particularly by the amount of insolation during the northern summer [7; 8]. The Asian summer monsoon system has also been shown to play a critical role [9]. However, these studies have used simplified models with a restricted spatial domain [8], or have only considered ENSO behaviour at 6 ka [7; 9].

This study aims to extend the spatial and temporal coverage of previous work by using a global climate model to explore the changes in the climate system over the past 8,000 years. By studying how the forcing "signal" translates into the observed "response", we aim to learn more about ENSO dynamics and, in particular, about the links between ENSO dynamics and global climate change.

2. Modelling experiments

The model used in this study is version 1.1 of the CSIRO Mk3L climate system model, a fully-coupled atmosphere-sea ice-ocean general circulation model [10]. The model comprises atmospheric and oceanic general circulation models, a dynamic-thermodynamic sea ice model and a land surface scheme with static vegetation. The atmosphere model has a horizontal resolution of $5.6^\circ \times 3.2^\circ$ and 18 vertical levels, while the ocean model has a horizontal resolution of $2.8^\circ \times 1.6^\circ$ and 21 vertical levels.

We conduct equilibrium simulations for 8, 7, 6, 5, 4, 3, 2, 1 and 0 ka. Each experiment differs *only* in that the Earth's orbital geometry is varied; otherwise the model is configured for pre-industrial conditions, with an atmospheric CO_2 concentration of 280 ppm. While this represents a somewhat idealised experimental design, these boundary conditions are nonetheless representative of the true conditions that prevailed during this period. Each simulation is integrated for 1,200 years, with the final 1,000 years being used for analysis.

We find that the model is broadly able to reproduce the trends in ENSO variability over the past 8,000 years (Figure 1). After an initial period of reduced variability, there is a gradual strengthening of ENSO variability after 5 ka, and a peak at 1 ka. Variability in the Niño 3.4 region is 16% weaker at 6 ka than at present. This is consistent with the range of uncertainty within the palaeoclimate record, which indicates a reduction in ENSO variability of 15–60% [11]. We therefore conclude that the model is a useful tool for studying the links between ENSO dynamics and global climate change.

3. ENSO dynamics

We now use the model simulations to follow the physical links within the climate system, as the forcing signal (i.e. the insolation changes) is translated into the observed response. Figure 2a shows the simulated June-July-August surface air temperature at 8 ka, expressed as an anomaly relative to the 0 ka simulation. The increased summer insolation at northern mid-latitudes gives rise to temperatures over land that are up to 4 K higher than at present. However, because the oceans have a much greater heat capacity than the land, the temperature changes over the ocean are smaller.

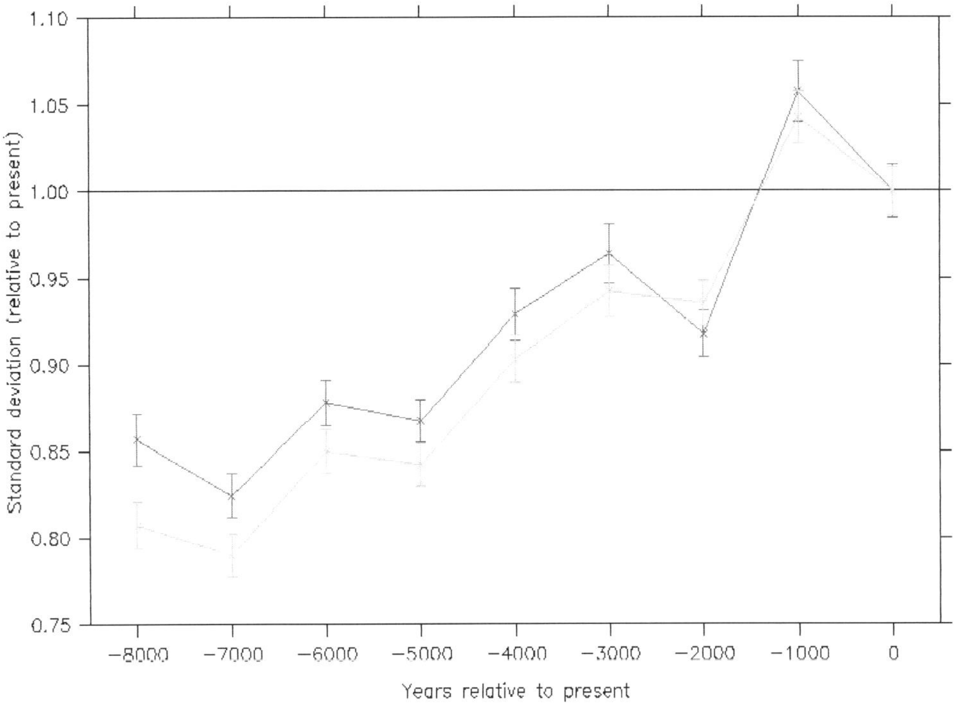

Figure 1. The standard deviation in the simulated monthly sea surface temperature anomalies for the Niño 3 (red) and Niño 3.4 (green) regions, normalised relative to the standard deviation for the present-day simulation. The 95% confidence intervals are shown.

The increased land-sea temperature contrast leads to an intensification of the Asian summer monsoon system, as can be seen in Figure 2b. The mean sea level pressure over the Eurasian and northern African landmasses is reduced by up to 6 hPa, with the enhanced deep convection drawing increased inflows of air from over the tropical Pacific and Indian oceans. The Walker Circulation over the Pacific Ocean is intensified, with an increase in the strength of the easterly trade winds in the central and western Pacific.

The intensification of the Walker Circulation changes the susceptibility of the coupled atmosphere-ocean system to the initiation of El Niño events. Figure 3a shows the propagation of westerly wind bursts within the 0 ka and 8 ka simulations. At 0 ka, most of the wind bursts propagate eastwards to the central Pacific, where they are able to perturb the easterly trade winds and trigger El Niño events. The peak variability in the wind stress occurs at about 160°W. At 8 ka, however, many of the wind bursts are confined to the western Pacific. Peak variability in the wind stresses now occurs at around 160°E, with only a few wind bursts breaking through to the central Pacific. We therefore find that the stronger easterly trade winds act as a barrier, blocking the propagation of westerly wind bursts and inhibiting the onset of El Niño events.

The effects upon ENSO variability are apparent from Figure 3b. At 0 ka, strong, regular El Niño events take place, with peak variability occurring in the central Pacific at around 150°W. At 8 ka, however, El Niño variability is much weaker, consistent with the palaeoclimate record.

The same mechanism is found to operate across all the simulations (not shown). As the summer insolation at northern mid-latitudes decreases from 8 ka to 0 ka, the Asian summer monsoon system and Walker Circulation weaken. The tropical Pacific therefore becomes increasingly susceptible to the initiation of El Niño events. While this mechanism explains the upward trend in ENSO variability over the past 8,000 years, however, it does not account

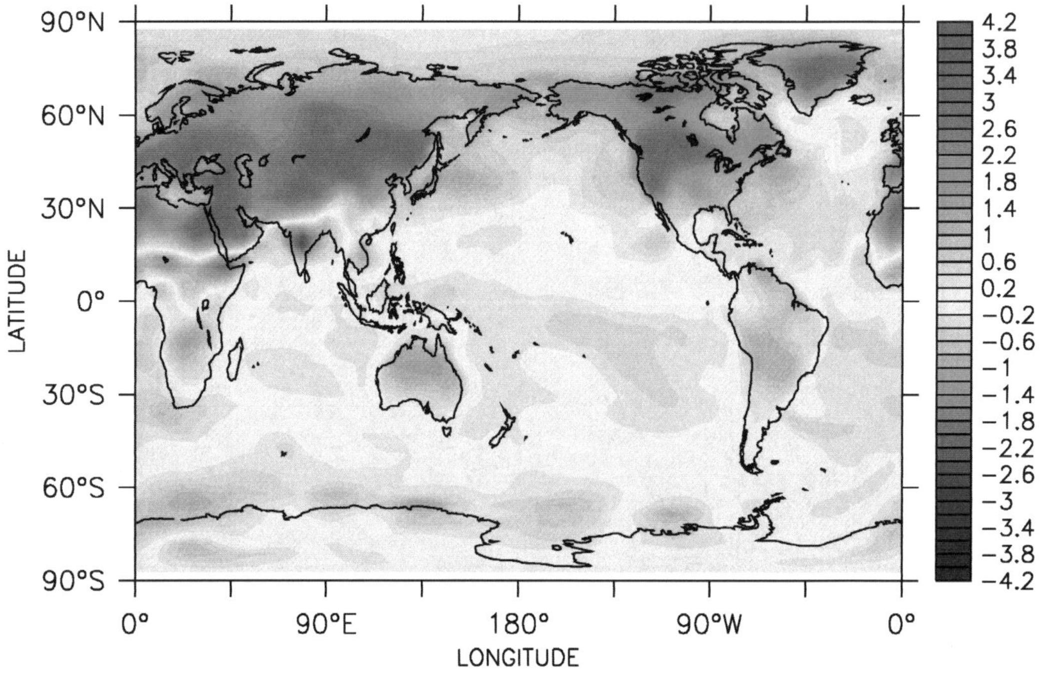

(a) Surface air temperature

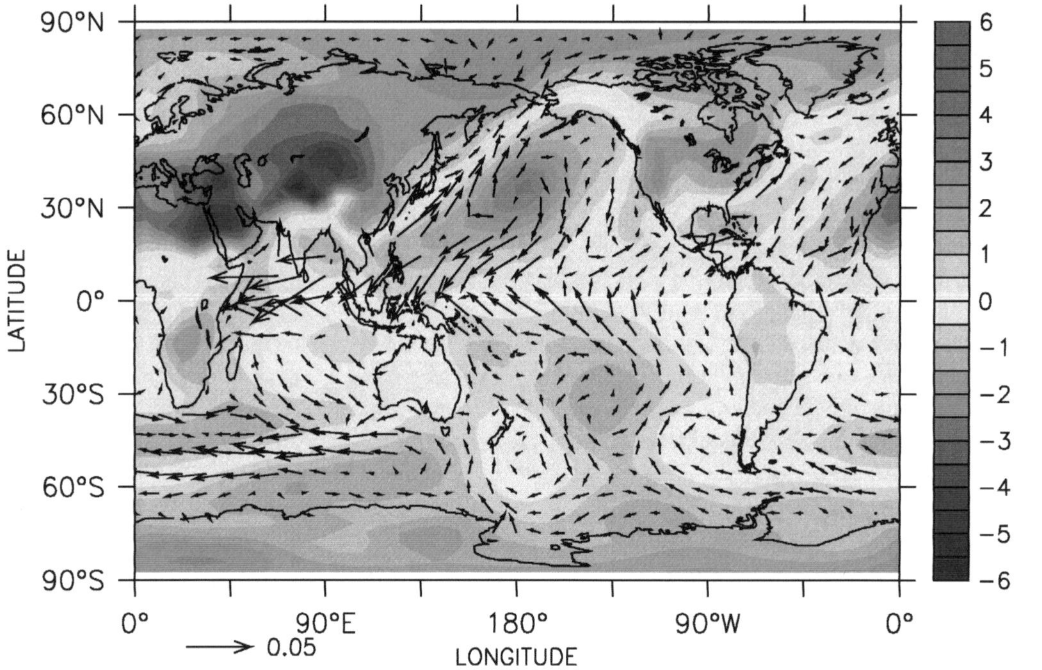

(b) Mean sea level pressure

Figure 2. June-July-August climatologies for the 8 ka simulation, expressed as anomalies relative to 0 ka: (a) surface air temperature (K), and (b) mean sea level pressure (hPa). In (b), the vectors show the surface wind stress anomalies over the ocean (Nm^{-2}).

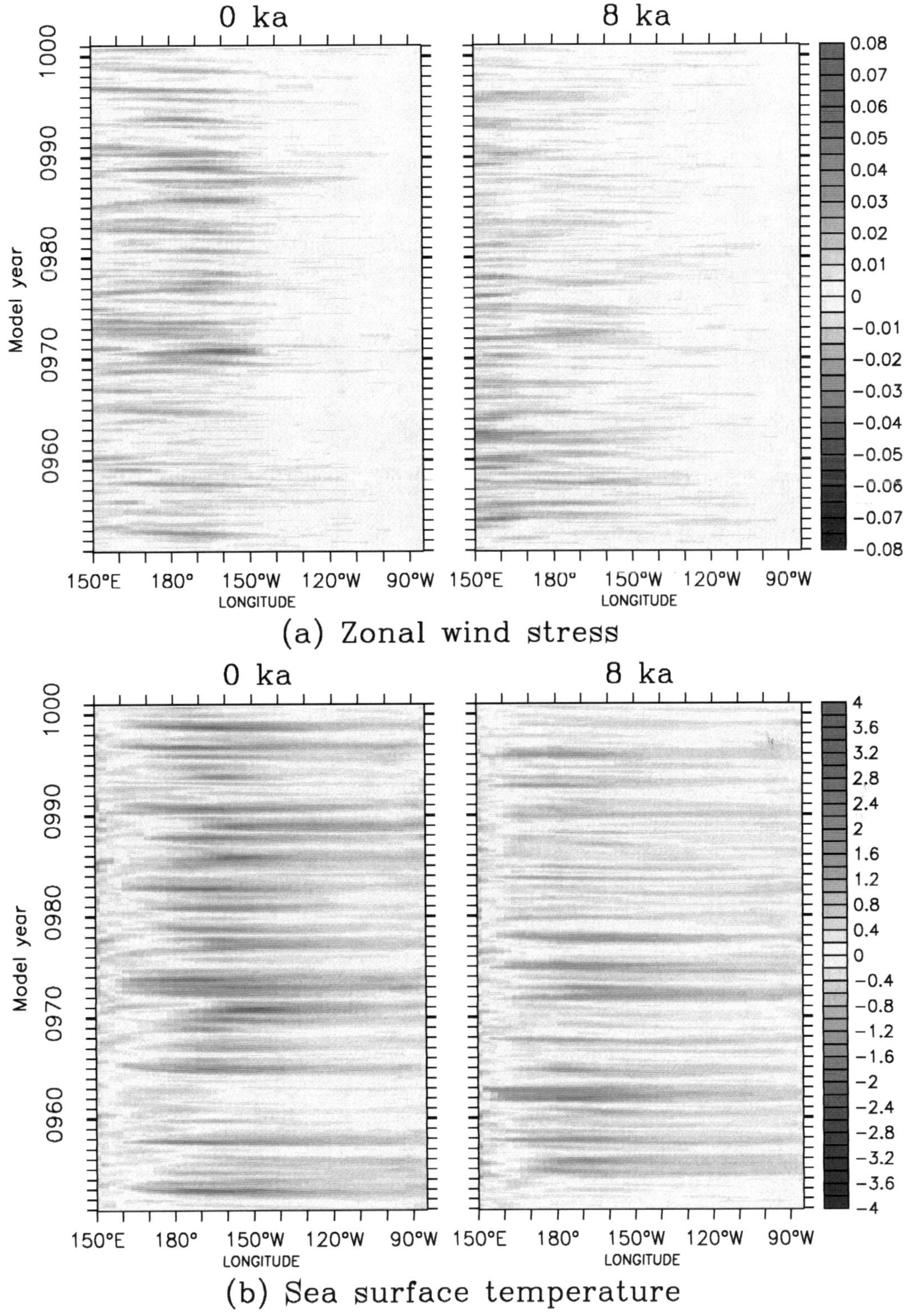

Figure 3. Monthly anomalies at the equator, relative to the mean seasonal cycle, for the final 50 years of the 0 ka and 8 ka simulations: (a) zonal wind stress (Nm^{-2}), and (b) sea surface temperature (K). The values shown are the five-month running mean.

for the simulated peak at 1 ka. Other mechanisms may therefore be at work.

Future studies should examine the changes in the magnitude, frequency and duration characteristics of El Niño events, and should investigate the effects of the changes in the atmospheric circulation upon ENSO modes. The current study only considers the role of the Earth's orbital geometry in driving changes in ENSO behaviour, and future work should therefore investigate the role of other factors, including volcanic activity and changes in solar luminosity and atmospheric greenhouse gases. The equilibrium response of CSIRO Mk3L to a doubling of the CO_2 concentration is a reduction of just 8% in the ENSO amplitude [12]. Changes in greenhouse gases may therefore have played only a minor role in the changes in ENSO behaviour over the past 8,000 years; however, this hypothesis should be thoroughly tested using multiple climate models.

4. Conclusions

By forcing a climate system model with orbitally-driven insolation changes, we have been able to reproduce the trends in ENSO variability over the past 8,000 years. Decreasing summer insolation over this period has resulted in a weakening of the Asian summer monsoon. This has reduced the stability of the background state of the tropical Pacific, creating conditions more favourable for El Niño development. However, other mechanisms may also be at work. A full understanding of the processes that drive changes in ENSO variability may be within grasp, but will require an approach that integrates the data, modelling and theory communities.

Acknowledgments

This work was supported by an award under the Merit Allocation Scheme on the NCI National Facility at the Australian National University. Travel support was received from the ARC Research Network for Earth System Science and the Australasian Quaternary Association. The authors wish to acknowledge use of the Ferret program for analysis and graphics in this paper (http://ferret.pmel.noaa.gov/Ferret/). The authors also wish to thank Jay Alder and Thorsten Kiefer for helpful comments on an earlier version of this manuscript.

References

[1] Philander S G 1990 *El Niño, La Niña, and the Southern Oscillation* (Academic Press)

[2] Rodbell D T, Seltzer G O, Anderson D M, Abbott M B, Enfield D B and Newman J H 1999 *Science* **283** 516–520

[3] Tudhope A W, Chilcott C P, McCulloch M J, Cook E R, Chappell J, Ellam R M, Lea D W, Lough J M and Shimmield G B 2001 *Science* **291** 1511–1517

[4] Moy C M, Seltzer G O, Rodbell D T and Anderson D M 2002 *Nature* **420** 162–165

[5] McGregor H V and Gagan M K 2004 *Geophysical Research Letters* **31** L11204

[6] Monnin E, Steig E J, Siegenthaler U, Kawamura K, Schwander J, Stauffer B, Stocker T F, Morse D L, Barnola J M, Bellier B, Raynaud D and Fischer H 2004 *Earth and Planetary Science Letters* **224** 45–54

[7] Liu Z, Kutzbach J and Wu L 2000 *Geophysical Research Letters* **27** 2265–2268

[8] Clement A C, Seager R and Cane M A 2000 *Paleoceanography* **15** 731–737

[9] Zheng W, Braconnot P, Guilyardi E, Merkel U and Yu Y 2008 *Climate Dynamics* **30** 745–762

[10] Phipps S J 2006 The CSIRO Mk3L climate system model Tech. Rep. 3 Antarctic Climate and Ecosystems Cooperative Research Centre Hobart, Tasmania, Australia

[11] Brown J, Collins M, Tudhope A W and Toniazzo T 2008 *Climate Dynamics* **30** 19–36

[12] Phipps S J *in preparation*

IOP Conf. Series: Earth and Environmental Science **9** (2010) 012011 doi:10.1088/1755-1315/9/1/012011

Could anthropogenic soil erosion have influenced Mediterranean vegetation distribution over the Holocene?

P M Collins[1], J O Kaplan and B A S Davis

Soil-Vegetation-Atmosphere Research Group ARVE, Institute of Environmental Science and Technology, Ecole Polytechnique Fédéral de Lausanne, Station 2, 1015 Lausanne, Switzerland

E-mail: pamela.collins@epfl.ch

Abstract. The circum-Mediterranean region is characterized by a strongly seasonal climate with rainy winters and intense summertime drought, steep topography, and a multi-millennial history of intensive human land use, all of which make its soils vulnerable to erosion. The historical and stratigraphic record documents severe and long-term soil erosion in several locations in the Mediterranean. A forest-to-scrub transition in Mediterranean vegetation between the mid-Holocene (6,000 yr BP) and the present is evident in the observational palaeorecord. Debate as to the causes of this shift is ongoing. This study seeks to test the sensitivity of large-scale vegetation patterns to changes in soil physical properties such as depth, content of coarse fragments, and organic matter content using the Mediterranean region as a case study. We find that simulated biomes are sensitive to changes in some soil physical properties at some locations, but that threshold values for soil change to affect vegetation are very high. Additional work is required to analyze the role that other soil physical properties, and climate change, played in influencing Holocene land cover change in the Mediterranean, and to improve model representations of relevant processes.

1. Introduction

Today, the lands bordering the Mediterranean Sea are largely characterized by coarse-textured shallow soils and sclerophyllous xeric vegetation [1]. Yet pollen records from the mid-Holocene (6,000 yr BP) suggest that several places currently dominated by xerophytic (drought-tolerant evergreen) vegetation once supported extensive deciduous forests [2, 3]. Evidence suggests that the northern limit for xerophytic woodland and scrub in southern Europe was located up to three degrees of latitude farther south as compared to its present position [4]. The debate about the extent [5] and causes [6-8] of this observed northward expansion of xerophytic scrubland vegetation in the Mediterranean region between the middle and late Holocene is ongoing. Long-term changes in regional climate may have been the main driver, as orbitally forced changes in atmospheric circulation resulted in alterations in temperature [9] and precipitation [10] patterns in space and time and affected seasonal moisture regimes. Alternatively, the influence of changing human land use patterns may have played a role, as deforestation, grazing, crop cultivation, and use of fire exerted strong anthropogenic selection pressures on natural vegetation communities [11, 12]. As the combined effects of anthropogenic disturbance and climate change can dramatically alter soil moisture content by promoting soil erosion

[1] To whom any correspondence should be addressed.

in uplands and sediment deposition in lowlands (for an overview, see [13]), our work attempts to contribute to this debate by posing the questions, "Do Mediterranean vegetation communities respond to changes in soil physical properties? If yes, how sensitive is the vegetation?"

Moisture availability often limits plant growth in seasonally arid environments such as the Mediterranean. Here, soils can play a prominent role in determining land cover by acting as moisture reservoirs during periods of drought [14]. Land cover and soil characteristics control the partitioning of precipitation into runoff and infiltration, and soil depth and texture influence the proportion of soil moisture lost to evaporation and drainage versus that retained in the soil matrix. Under identical climatic conditions, differences in soil depth and texture in adjacent patches can cause marked differences in vegetation density [15, 16] and species distribution [17].

Though modern Mediterranean upland soils tend to be coarse, shallow, and poor in organic matter content, evidence suggests that this was not always the case. The Global Assessment of Soil Degradation classifies the soils of much of the circum-Mediterranean region as moderately or severely degraded (approximately 97,000 km^2 in Spain, 70,000 km^2 in Italy, and 40,000 km^2 in Greece [18]), with regions of very severe degradation [18, 19], implying that conditions were once quite different. During the classical periods of Greek and Roman civilization (ca. 2-3 ka BP), currently degraded landscapes such as those near ancient Carthage (Mediterranean north Africa) were once productive enough to feed cities and empires [20].

The combination of steep terrain, strongly seasonal and episodic precipitation, and a multi-millennial history of continuous and often intensive human land use make the Mediterranean region particularly prone to upland erosion and lowland sedimentation. Alluvial stratigraphic records show several distinct events of elevated soil erosion throughout the Mediterranean during the Holocene; they vary in both duration and severity by location and may be correlated with fluctuations in both climate and regional civilizations [21, 22, 23] (for a review, see [13]).

Research programs such as ARCHAEOMEDES [24] and MEDALUS (Mediterranean Desertification and Land Use) [25] had the aim of qualitatively understanding land degradation in the Mediterranean on a regional or site-by-site basis in historical times and the present day, respectively. As yet, however, there has been no attempt to investigate the impact of changing soil depth and texture on vegetation cover across the Mediterranean region in a systematic, quantitative manner. This study does not attempt to explain the changes in Mediterranean land cover over the Holocene observed in the pollen record, nor does it intend to make comparisons between pollen-reconstructed and model-calculated vegetation cover. Rather, it contributes to the ongoing debate about the causes of Mediterranean land cover change over the Holocene by testing the sensitivity of Mediterranean vegetation to changes in soil physical properties using a simple modeling approach.

2. Methods

To simplify our analysis and focus on those areas where we might expect changes in soil physical properties to have influenced vegetation cover in the past, we used the paleoecological record to identify a number of sites where vegetation changed from forest to non-forest types from the mid-Holocene to present. At 34 sites from across the Mediterranean region, we used the SoilMod soil physical model and the BIOME1 equilibrium vegetation model [14] to quantify the sensitivity of biome-scale vegetation to changing soil physical properties. The sites where we ran our model are broadly representative of the range of Mediterranean climate and soil characteristics, from mountains to coastlines and from semi-mesic to xeric environments. A detailed description of our methods follows.

2.1. Site description

The Mediterranean region lies roughly between 30°N and 40°N latitude and 10°E and 40°W longitude and has been defined as the region in which olive trees can be cultivated, the northern boundary of which is roughly determined by the 5°C January isotherm [26]. It is a tectonically young landscape with several steep mountain ranges. The climate is seasonal, with summers characterized by clear skies and drought due to the falling limb of the Hadley cell and winter weather dominated by the prevailing westerlies that deliver cyclonic precipitation with high spatial and temporal heterogeneity due to the varied topography. Average annual precipitation ranges from approximately 100 to 900 mm (two-thirds of which fall between September and March [1]), and the mean annual temperature is 15 ± 5 °C [4].

2.2. Data sources and study sites

To identify specific sites that experienced a change in land cover between the mid-Holocene and the present, this study uses a published pollen-based reconstruction of Mediterranean land cover classified into 12 biomes (following [27]) at 122 sites for both the present day and the mid-Holocene [4]. We classified the biome assignment for each site for both time periods as forest or non-forest, according to each biome's annual moisture requirement index α (Priestly-Taylor coefficient, or ratio of actual to potential evapotranspiration, AET/PET) as described in [14] (forest: $1 \geq \alpha \geq 0.65$; non-forest: $0.65 > \alpha \geq 0$). Thirty-four sites shifted from forest to non-forest between 6 ka BP and the present (figure 1), 3 sites did the reverse, and 85 sites did not cross the $\alpha = 0.65$ threshold in either direction. For each of the 34 sites that shifted from forest to non-forest, we extracted late 20th century soil data (depth, particle size distribution, organic matter content, percent of coarse fragments) at the 1 km gridcell nearest to each site from the Harmonized World Soil Database (HWSD) [28] and late 20th century mean climatology (temperature and precipitation values as used in [14]) from the WorldClim database [29]. The pollen data are not used again in this study after this site-selection step.

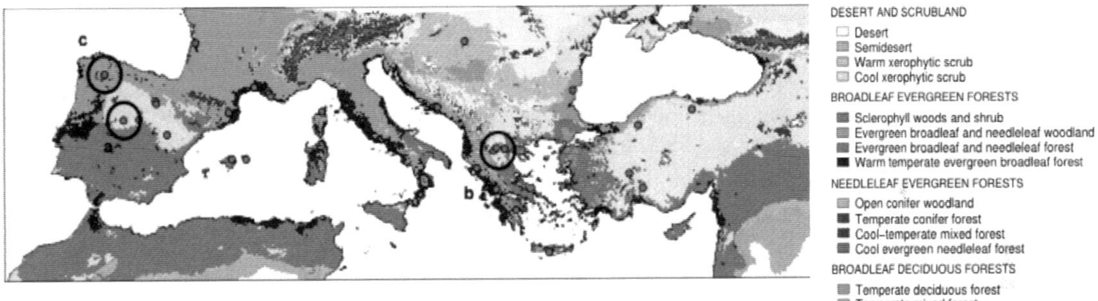

Figure 1. Potential natural vegetation of the Mediterranean study area simulated by the BIOME1 model. Each red circle represents a sediment core where pollen data indicate a forest biome at 6 ka BP and a non-forest biome in the present day [2]. Results for circled sites a, b, and c are presented in figures 2 and 3.

2.3. Models and calibration

We employed two models in this study. First, we used a soil physical model (SoilMod) driven by the HWSD parameters listed in section 2.2. This model consists of a series of pedotransfer functions [30-34] and calculates soil bulk density, water holding capacity (defined as the difference between field capacity, or water remaining in the soil under -33 kPa water potential [Ψ], and wilting point, or water remaining in the soil at Ψ -1,500 kPa), and saturated hydraulic conductivity for the region surrounding each of the 34 sites analyzed.

Next, we used a global biogeography model (BIOME1) driven by WorldClim climate data [29] and the output from SoilMod (see above). For each gridcell or data point, BIOME1 uses these data to determine the presence or absence of each of 13 distinct plant functional types (PFTs) based on their physiological requirements, namely minimum and maximum thresholds of temperature and moisture availability [14]. The latter is determined by the moisture availability index α, calculated as the ratio of actual to potential evapotranspiration (AET/PET) [14, 35]. AET is proportional to soil water content in a single-layer "bucket" type soil, which is a function of soil water holding capacity (as calculated by

IOP Conf. Series: Earth and Environmental Science **9** (2010) 012011 doi:10.1088/1755-1315/9/1/012011

SoilMod) and precipitation (from WorldClim). PET is calculated using the equations in [35] with temperature data from WorldClim and % sunshine data from the CRU CL2.00 climatology [36]. Manipulating the soil physical properties for a given location changes the size of the soil "bucket" and therefore leads to a change in the calculated α value for that location, which in turn may alter the location's suitability for a given PFT. Finally, a unique biome name is assigned to each combination of present and absent PFTs (figure 1; PFTs as in [14]).

To calibrate BIOME1, we used it to generate a gridded map of potential natural vegetation represented as biomes (figure 1) and then refined our model parameterizations until the map output satisfactorily matched the modern potential vegetation distribution for the Mediterranean [37].

2.4. Sensitivity tests

As depth and content of coarse fragments (% v) both influence the total volume of fine soil material that is available to retain water, and soil organic matter content (% w) affects a mineral soil's ability to retain water, these were the three soil physical parameters we chose to manipulate. Changing these three parameters alters the resulting bulk density, water holding capacity, and saturated hydraulic conductivity as calculated by SoilMod, which in turn modifies the α value used by BIOME1 to calculate the biome for each combination of parameters. Assuming that soil degradation would decrease depth and organic matter content and increase the percent of coarse fragments, we explored a wide range of parameter space that could represent both "degraded" and "improved" conditions relative to present-day soil conditions.

First, we varied the simulated depth of all soils from 0.1 m to 2.0 m and calculated soil water holding capacity using SoilMod, keeping all other soil characteristics constant. Then, we held depth constant at HWSD (present-day) values and simultaneously varied simulated soil organic matter from 0% to 20% (w) and content of coarse fragments (lithic material > 2 mm in diameter) from 100% to 0% (v) for all test sites, again using SoilMod. For both sets of experiments, we ran the new calculated soil bulk density, water holding capacity, and conductivity values through BIOME1 and checked the resulting simulated biome to see if a "degradation" or "improvement" in soil physical parameters under constant climatic conditions led to a change in the calculated biome assignment at any of our test sites.

3. Results

Of the 34 sites analyzed, the simulated biome was insensitive to any "improvement" in soil physical properties at all but three sites (figures 1, 2, 3). Each of these three sites was sensitive to a decrease in the content of coarse fragments, but only two were sensitive to an increase in depth, and none responded substantially to an increase in soil organic matter (figures 2, 3). Sensitivity was expressed by a change in the calculated biome for a site, e.g. from drought-tolerant open conifer woodland to drought-sensitive temperate deciduous forest (figures 2, 3).

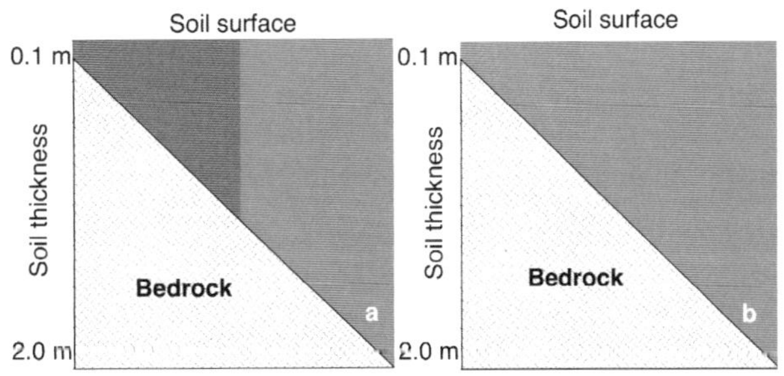

Figure 2. Increasing soil depth from 0.1 m to 2.0 m led to a biome shift at two of our 34 test sites. Letters correspond to circled sites in figure 1. Colors indicate biome types. Green: temperate deciduous forest; blue: evergreen broadleaf and needleleaf forest; rose: evergreen broadleaf and needleleaf woodland. For biome descriptions, see [14].

For this study, the majority of sites appear insensitive to increases in soil depth and organic matter and decreases in the content of coarse fragments. There are three potential explanations for this outcome: 1.) Changes may have occurred in the relative percentages of PFTs present, or even in levels of net primary productivity, but the resolution of our biome-level analysis was too coarse to pick up these subtleties; 2.) Soil water holding capacity is more sensitive to changes in soil texture (which we have not yet tested) than to soil depth, stoniness, or organic content; 3.) We tested the correct parameters and analyzed the output at the appropriate level of complexity, but in fact Mediterranean vegetation is insensitive to changes in soil physical properties.

The 3 sites that did respond to an "improvement" of soil physical properties were likely already near the threshold between two different biome classes (figure 1). However, all 34 sites responded to a "degradation" of soil physical properties; when soils approached a threshold level of 95%-97% (v) coarse fragments, biomes shifted uniformly to desert conditions.

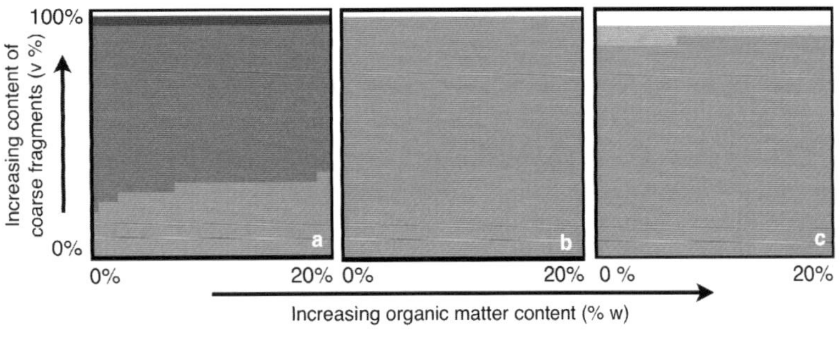

Figure 3. Changing soil organic matter content and content of coarse fragments caused a shift in simulated biome at three of our 34 test sites. Letters correspond to circled sites in figure 1. Change in color indicates biome shift. Pale Yellow: desert; dark green: temperate conifer forest; pink: open conifer woodland. For other colors, see figure 2.

Rather than analyzing the entire circum-Mediterranean region in a gridded manner, this study works with 34 individual points distributed across the entire region that are assumed to be representative of the full range of Mediterranean climate and soil characteristics. This was done to simplify our approach; we acknowledge that future work would benefit greatly from analyzing the entire gridded region and intend to do so in the future. The biome scheme used in BIOME1 is not analogous to the biome scheme used with the pollen data, making direct comparisons between biomes from model output and biome assignments from pollen data impossible. This discrepancy will be remedied in future work. For this study, we are more interested in whether or not a shift from one biome type to another occurs in response to changes in soil physical properties; the actual biome assignments are of secondary interest.

4. Discussion and Conclusions

Our results suggest that at least under certain circumstances, "improving" soils by increasing soil depth and organic matter content and/or decreasing the content of coarse fragments can lead to the simulation of more moisture-loving biome types (i.e. a shift from evergreen to deciduous vegetation) without requiring a change in climate, and reversing the direction of these changes can lead to the simulation of desert. However, the threshold levels for these changes are quite high and suggest that further study is required to improve our model representations of the processes involved in soil-vegetation-climate interactions.

This study attempts to quantify the sensitivity of vegetation under Mediterranean conditions to soil thickness, stoniness, and organic matter content. Though all 34 sites analyzed demonstrate a clear transition from forest to non-forest biomes between the mid-Holocene and the present in the pollen record, only three respond to changes in modeled soil stoniness and organic matter content (figure 3), and just two of these show a change in biome in response to changing soil depth (figure 2). These results could suggest that Mediterranean vegetation is relatively insensitive to changes in soil physical properties. However, only a small number of soil parameters were tested for a limited number of locations, and more complex analysis over a wider area may reveal greater sensitivity.

We will follow up on this preliminary work with a more in-depth analysis of the environmental factors relevant at our study sites, and we will extend our analysis to additional sites. We will expand our experimental design to analyze soil texture (relative proportions of sand, silt, and clay), we will use an alternative soils dataset [38, 39] that provides more detailed information on soil depth at the present day, and we will analyze vegetation response at the levels of net primary productivity and plant functional types. We will also incorporate potential soil-climate feedbacks and interactions in future work.

The question of vegetation sensitivity to changes in soil physical characteristics is not merely academic. It may have serious implications for the concepts of sustainable land use and resilience of land to abuse and degradation. How much soil can be lost (or gained) before a region's natural vegetation community shifts from one stable state to another? Is the process reversible? Do thresholds differ between ecosystems? Though much work has been done on the annual to decadal detrimental effects of soil erosion on agricultural productivity [40-43], little quantitative data is available on how centuries or millennia of human land use alter the land's capacity to support natural vegetation or human activities. The circum-Mediterranean region is a drought-limited, seasonal environment prone to erosion that has been heavily used by humans for millennia. In many places, Mediterranean soils have experienced dramatic erosion since the mid-Holocene. The debate about what caused the observed shift in vegetation from forest to non-forest between the mid-Holocene and the present in many areas of the Mediterranean is still open. Any attempt to understand the dynamics of vegetation over long timescales should not ignore the potential impact of changes in soil physical properties under human, or even natural, influences.

Acknowledgements

Funding for this work was provided by grants from the Swiss National Science Foundation (PP0022_119049) and the Italian Ministry for Research and Education (FIRB RBID08LNFJ) for the Research Project CASTANEA. The authors also thank Dr. José Carrión and Dr. Paul Henne for their helpful comments on the manuscript of this paper.

References

[1] Breckle S-W 2002 *Walter's Vegetation of the Earth: The Ecological Systems of the Geo-Biosphere* (Berlin: Springer)
[2] Bottema S 1980 *Rev. Palaeobot. Palynol.* **31** 193-217
[3] Jahns S 1993 *Veget. Hist. Archaeobot.* **2** 187-203
[4] Roberts N, Stevenson T, Davis B, Cheddadi R, Brewster S and Rosen A 2004 in Battarbee R W et al. (eds) *Past Climate Variability through Europe and Africa* (Dordrecht: Springer)
[5] Grove A T and Rackham O 2003 *The Nature of Mediterranean Europe: An Ecological History* (New Haven: Yale University Press)
[6] de Beaulieu J-L, Miras Y, Andrieu-Ponel V and Guiter F 2005 *Plant Biosystems* **139(2)** 114-26
[7] Colombaroli D, Tinner W, van Leeuwen J, Noti R, Vescovi E, Vannière B, Magny M, Schmidt R and Bugmann H 2009 *Journal of Biogeography* **36** 314-26
[8] Jalut G, Dedoubat J J, Fontugne and Otto T 2009 *Quaternary International* **200** 4-18
[9] Davis B A S, Brewer S, Stevenson A C, Guiot J and Data Contributors 2003 *Quaternary Science Reviews* **22** 1701-16
[10] Harrison S P and Digerfeldt G 1993 *Quaternary Science Reviews* **12** 233-48
[11] Carrión J S, Fuentes N, González-Sampériz P, Sánchez Quirante L, Finlayson J C, Fernández S and Andrade A 2007 *Quaternary Science Reviews* **26** 1455-75
[12] Carrión J S, Fernández S, Jiménez-Moreno G, Fauquette S, Gil-Romera G, González-Sampériz P and Finlayson C 2009 *Journal of Arid Environments* **in press** 1-6
[13] Bintliff J 2002 *World Archaeology* **33(3)** 417-35
[14] Prentice I C, Cramer W, Harrison S P, Leemans R, Monserud R A and Solomon A M 1992 *Journal of Biogeography* **19** 117-34
[15] Henkin Z, Seligman N G, Kafkafi U and Prinz D 1998 *Plant and Soil* **202** 317-26
[16] Kosmas C, Danalatos N G and Gerontidis St 2000 *Catena* **40** 3-17
[17] Theodoropoulos T, Reif A and Athanasiadis N 1995 *Bot Helv* **105** 37-54
[18] National Soil Degradation Maps [homepage on the Internet] FAO/UNEP/ISRIC 1990 [updated 2005 December 12; cited 2009 September 17]. Available from: http://www.fao.org/landandwater/agll/glasod/glasodmaps.jsp?country=NER&search=Display+map+%21
[19] Bridges E M and Oldeman L R 1999 *Arid Soil Research and Rehabilitation* **13** 319-25
[20] Montgomery D 2007 *Dirt: The erosion of civilizations* (Berkeley: University of California Press)
[21] Shiel R S 1999 in Leveau P et al. (eds) Environmental Reconstruction in Mediterranean Landscape Archaeology (Oxbow)
[22] Bruckner H 1990 in Bottema et al. (eds) *Man's Role in the Shaping of the Eastern Mediterranean Landscape* (Rotterdam: Balkema)
[23] OP-EUR14850 1994 (Luxembourg: European Comission)
[24] van der Leeuw S E and the ARCHAEOMEDES research team 2005 *C. R. Geoscience* **337** 9-27
[25] Kosmas C S 1995 Field site: Spata, Greece, MEDALUS II-Project 1, Basic Field Program, Final Report Covering the Period "1 Janary 1991 to 1 September 1995"
[26] Rother K 1993 *Der Mittelmeerraum* (Stuttgart: Teubner)
[27] Peyron O, Guiot J, Cheddadi R, Tarasov P, Reille M, de Beaulieu J-L, Bottema S and Andrieu V 1998 *Quaternary Research* **49** 183-96
[28] FAO/IIASA/ISRIC/ISS-CAS/JRC 2009 *Harmonized World Soil Database v. 1.1* (Rome: FAO and Laxenburg, Austria: IIASA)
[29] Hijmans R J, Cameron S E, Parra J L, Jones P G and Jarvis A 2005 *International Journal of Climatology* **25** 1965-78
[30] Brooks R H and Corey A T 1964 *Hydrology Papers* (Fort Collins, CO: Colorado State

University)
[31] Clapp R B and Hornberger G M 1978 *Water Resources Research* **14(4)** 601-4
[32] Van Genuchten M T 1980 *Soil Sci. Soc. Am. J.,* **4** 892-98
[33] Brakensiek D L and Rawls W J 1994 *Catena* **23** 99-110
[34] Saxton K E and Rawls W J 2006 *Soil Sci. Soc. Am. J.* **70** 1569-78
[35] Prentice I C, Sykes M T and Cramer W 1993 *Ecological Modelling* **65** 51-70
[36] New M, Hulme M and Jones P 1999 *Journal of Climate* **12** 829-856
[37] Bohn U, Gollub G and Hettwer C 2000 *Karte der natürlichen Vegetation Europas 1 : 2 500 000* (Bonn-Bad Godesberg, Germany: Bundesamt für Naturschutz)
[38] Webb R S, Rosenzweig C E and Levine E R 1991 A Global Data Set of Soil Particle Size Properties *NASA Technical Memorandum 4286* (NASA)
[39] Webb R S, Rosenzweig C E and Levine E R 1993 *Global Biogeochemical Cycles* **7(1)** 97-108
[40] De la Rosa D, Moreno J A, Mayol F and Bonsón T 2000 *Agriculture, Ecosystems & Environment* **81(3)** 179-90
[41] den Biggelaar C, Rattan L, Wiebe K and Breneman V 2003 *Advances in Agronomy,* **81** 1-48
[42] Larney F J, Janzen H H, Olson B M and Olson 2009 *Soil and Tillage Research* **103(1)** 73-83
[43] Sui Y, Liu X, Jin J, Zhang S, Zhang X, Herbert S J and Ding G 2009 *Field Crops Research* **111(3)** 276-83

Sedimentation adjacent to naturally eroding and breakwater-protected shorelines in Chesapeake Bay

C M Palinkas*, E W Koch and N Barth

University of Maryland Center for Environmental Science, Horn Point Laboratory, Cambridge, MD, USA

*cpalinkas@umces.edu

Abstract. Sediment characteristics, especially grain size and organic content, in nearshore Chesapeake Bay environments show significant temporal and spatial variability. This can impact benthic organisms, particularly submerged aquatic vegetation (SAV), which are important components of the ecosystem. In order to better understand how these changes are reflected in the stratigraphic record, the radiochemical and textural properties of sediment at four sites are examined. Fine and organic material are observed to increase at some nearshore locations, whereas others have experienced a shift toward lower-organic, coarser sediments. These changes are likely related to local variations in sedimentary processes. Other, more recent, perturbations are due to breakwater construction, which can trap fine and organic material in the protected area. Accumulation rates inshore of the breakwater are ~2-4 times higher than in adjacent exposed locations, and this change is coincident with breakwater construction. Thus, because sedimentary processes vary according to physical setting, local trends must be discerned to determine whether a given site may be suitable for SAV restoration.

1. Introduction

The nearshore environment is dynamic in nature. Sedimentary [1, 34], physical and chemical processes (e.g., freshwater input, wave energy, light penetration, nutrient cycling, etc.) change over time and space. These nearshore processes are important in determining the quality of benthic habitats [14, 32, 33, 35] occupied by valuable living resources, including shellfish and submersed aquatic vegetation (SAV). The role of sedimentary processes in habitat requirements is better understood for some species than for others. For example, while it is well established that oysters require a hard substrate to settle and grow [15], only recently has the importance of substrate for SAV been highlighted [17]. The proximity of SAV habitats to shore heightens the potential influence of anthropogenic activities such as shoreline hardening (i.e. constriction of breakwaters, groins, rip rap) on nearshore sedimentary processes [21, 22].

Previous studies of Chesapeake Bay sedimentation have revealed a complex history of human influence on sedimentary processes. Multiple changes to the Bay and its watershed (e.g., land-use changes since European settlement, increases in nutrient influxes) have altered ecosystem processes such as sedimentation rates, turbidity, eutrophication, and anoxia [5, and references therein]. In recent years, these changes have included modifications to the nearshore and shoreline via coastal structures (e.g., breakwaters, groins, jetties, etc.) and hardened shorelines (e.g., bulkheads, rip rap). These protection structures typically arise in response to high coastal erosion rates in the region – up to 1.8

m/y at our study sites [23, 24] – due to a combination of long-term local sea-level rise, storm-induced high sea level, and large waves [11]. Coastal structures alter local hydrodynamics, and thus also the sedimentary environment, which could ultimately affect benthic habitat quality [22]. However, relatively few previous studies have specifically targeted the shallow, nearshore areas potentially habitable by SAV (subtidal waters, <2 m), though there are notable exceptions [2, 6].

SAV is an important component of healthy ecosystems, providing habitat and food for many organisms [12, 20]. In Chesapeake Bay, SAV has experienced significant declines since the 1970s [27, 28], primarily due to water quality impairment by excess suspended sediments and dissolved nutrients that reduce light availability [9]. It is likely that other, sometimes co-varying, factors also affect SAV distribution and growth, including temperature, salinity, waves, sediment type and accumulation rates [8, 17]; these alternate factors can be as important as light in some systems [19]. For example, in Chincoteague Bay, VA, combining substrate (<35% silt and clay) and light (depth <2 m) requirements accurately predicts SAV distributions [16]. Thus, our study is designed not only to assess patterns of nearshore sedimentation, but also to inform SAV restoration efforts, which have met with mixed results [18, 26]. Thus, the goal of the present study is to better understand and compare temporal variations in sediment characteristics, particularly grain size and organic content, among nearshore Chesapeake Bay environments. We focus on changes in grain-size and organic-content trends occurring over the last <~100 y, relating both to coastal structures (i.e., breakwaters) and natural processes. Grain size and organic content have been chosen as proxies of potential suitability for SAV, as relatively well-defined ranges have been observed in the Bay, typically 6-10% fine (mud) material and 1.0-5.3% organic material [3]. In addition, the modern time period coincides with direct observations of SAV in the Bay (mid-1960s to present), though explicit comparisons of sediment properties and SAV distributions are left for future studies.

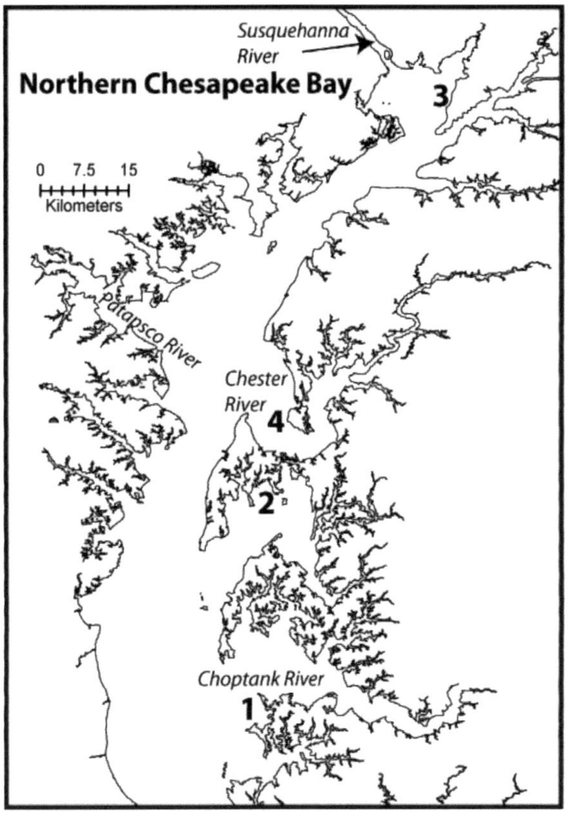

Figure 1. Core locations in Trippe Bay (1), Crab Alley Bay (2), Elk Neck (3), and Eastern Neck (4) in middle and upper Chesapeake Bay.

2. Materials and methods

Two vibracores (~3 m long) were collected at each of four study sites in 2008 (Figs. 1, 2). In Trippe and Crab Alley Bays, these cores were located within the area typically covered by SAV, as determined by weighted overlays of aerial photography data in ArcGIS [31]. For Elk and Eastern Necks, one core was placed behind (landward of) the breakwater and one was adjacent to it at a similar water depth. Because vibracores disturb the upper few centimeters of the sediment surface, companion push cores (~30 cm long) were also collected. Breakwater ages were determined via aerial photographs [28, prior annual reports 1978-present]. After collection, all cores were returned to the lab and frozen until immediately prior to analysis for grain size, organic content, and the naturally occurring radioisotope ^{210}Pb. Cores were subsectioned into 1- and 2-cm increments. Samples for grain size were wet sieved at 4 phi (64 μm) to separate the mud and sand components. The mud component was then dispersed with sodium metaphosphate and placed in an ultrasonic bath before analysis with a Sedigraph 5120.

IOP Conf. Series: Earth and Environmental Science **9** (2010) 012012 doi:10.1088/1755-1315/9/1/012012

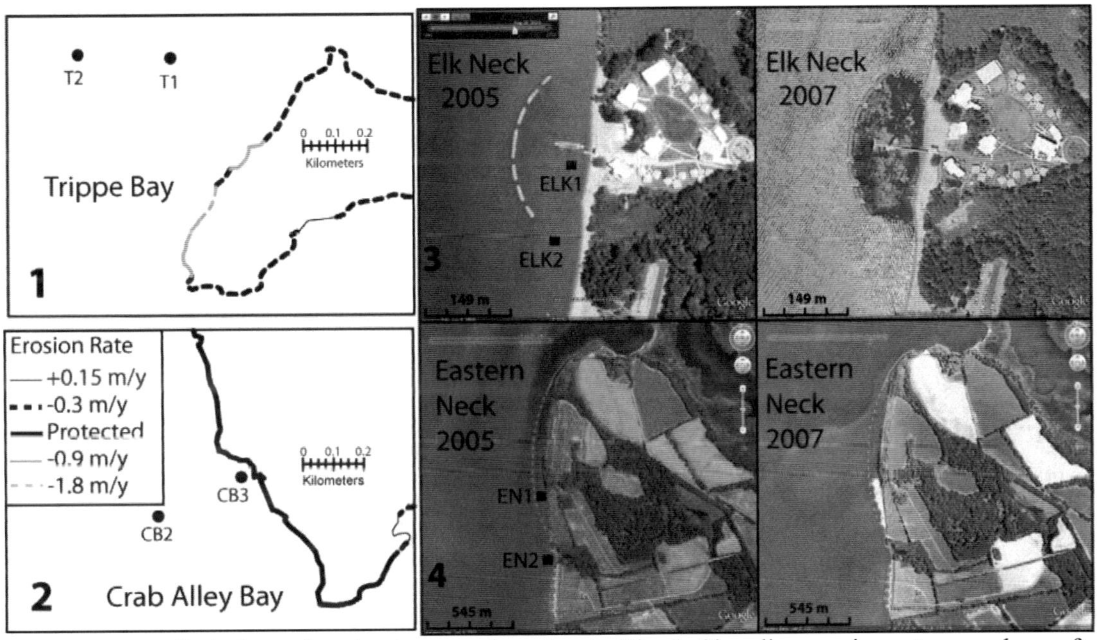

Figure 2. Detailed view of study sites, numbered as in Fig. 1. Shoreline erosion rates are shown for Trippe and Crab Alley Bays (right); positive and negative values denote accretion and erosion, respectively [2]. 2005 and 2007 Google Earth images show the footprint of breakwaters at Elk and Eastern Necks (left), as well as local environmental conditions. Dark patches in the 2007 Elk Neck image are likely SAV that are absent in 2005, suggesting an initial benefit of breakwaters to SAV through reduction of physical energy. At Eastern Neck, a sand deposit is present near site EN2 in 2007 but not 2005.

The sand component was dry sieved from 1 to 4 phi (500-64 μm) at ¼-phi increments. For the purposes of this study, sediment >1 phi was classified as coarse sand. Both mud and sand data were then combined to calculate the median diameter. Diameters are reported in phi units, facilitating statistical computations and graphical presentation (phi = $-\log_2$(particle diameter, mm)). Organic content was measured via combustion in a muffle furnace at 450°C for 4 hours [10]. ^{210}Pb (half-life 22.3 y) analyses were performed by alpha spectroscopy, following the procedures of [30]. The slope of the ln ^{210}Pb decrease with depth in each core was fit to a linear regression line and used to calculate an accumulation rate. ^{210}Pb can be detected for ~4-5 half-lives, therefore excess activity is present in sediment <~100-y old. ^{210}Pb-derived accumulation rates were verified by ^{137}Cs geochronology via gamma spectroscopy [30].

3. Results and discussion

3.1 Historical changes in sedimentation

Cores in Trippe and Crab Alley Bays exemplify two different observed historical changes. In Trippe Bay, sediment shifts at ~150 cm from more organic, fine sediment deeper in the core to lower organic, coarse sediment shallower in the core (Fig. 3a). In the deep section, sediment has an average median diameter of 4.4 phi (46.9 μm) and organic content of 2.8%. In the shallower section, the average median diameter is 3.4 phi (94.5 μm) and organic content is 1.2%. This change, occurring >~100 y ago (below the penetration depth of excess ^{210}Pb), likely relates to historical changes in the sedimentary environment. Shoreline changes in this region can be relatively rapid; present erosion rates are 0.3-1.8 m/y (see Fig. 1). If the present accumulation rate is extrapolated, this transition occurs in the mid-1800s, when the shoreline was ~100 m farther seaward [23, 24], and thus the depositional regime was likely different.

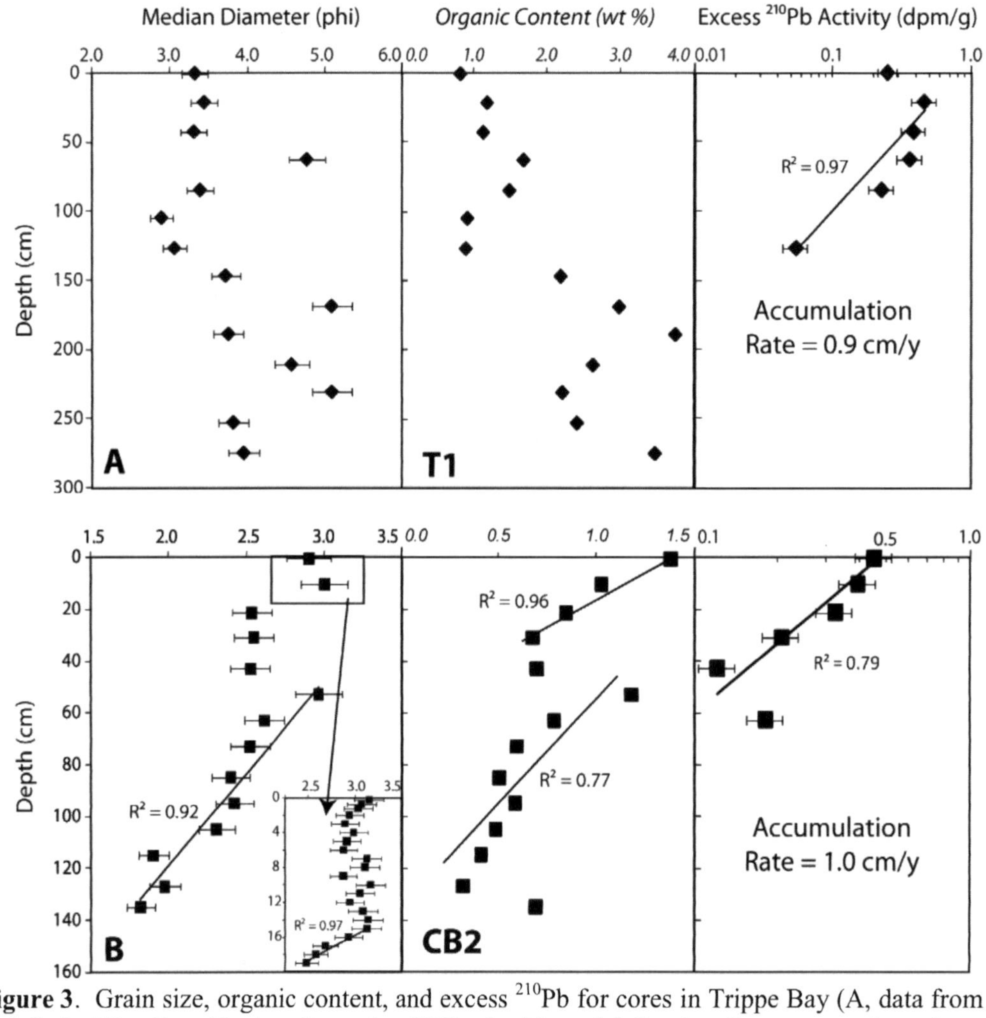

Figure 3. Grain size, organic content, and excess ^{210}Pb for cores in Trippe Bay (A, data from site T1) and Crab Alley Bay (B, data from site CB2). In this and following figures, error bars for grain size represent the average error observed in replicate cores at each location (typically ~5%), and error bars for excess ^{210}Pb represent the counting error, which is the largest source of error in determining activities. Note that x-axis values vary among plots. (B, inset). Grain-size data from the push-core at CB2, which provides a more detailed view of surficial sediment character, corresponding to the boxed area of the vibracore profile. Regression lines for the upper and lower portions of the sedimentary record at this site have different slopes. See text for details.

Crab Alley Bay exemplifies nearshore fining, similar to that observed in deeper waters of the Chesapeake Bay [7]. In this core, sediment fines and becomes more organic upward (Fig. 3b). Basal values for the median diameter and organic content are 1.8 phi (283.2 μm) and 0.7%, respectively, fining upward to surficial values of 2.9 phi (134.0 μm) and 1.4%. The general trend of increasing fine and organic content upward occurs throughout the entire core, but a change in the slope of increase occurs at 54 cm, below a ~20-cm relatively coarse, lower-organic layer. In the upper ~20 cm, sediment fines from 2.5 phi (178.9 μm) at 19 cm to 3.0 phi (123.0 μm) at 14 cm (Fig. 3b, inset). This change occurs at a rate of 0.20 phi/cm, an order of magnitude higher than the rate below 54 cm (i.e., the historical rate) of 0.01 phi/cm. Organic content also increases more rapidly above 54 cm (i.e., recent years), 0.02 %/cm, than below 54 cm, 0.01 %/cm. In the main stem, similar changes in sedimentary character have been related to land-use changes (i.e., agricultural clearance and deforestation) since European settlement [7], which is also likely the case in the nearshore. Overprinted on this long-term trend are potential changes related to shoreline hardening in this region (corresponding to the "protected" shoreline in Fig. 1). Shoreline hardening limits the exchange of sand between the shoreline and the nearshore, resulting in finer sediments dominating the record. If the sedimentation rate is assumed to be constant throughout, then this indicates that the fine and organic content of sediment has increased more rapidly in recent years. This could be due to a combination of shoreline hardening and continuation of the longer-term trend. Alternatively, a decrease in sedimentation rate and constant fine and organic content would produce the same trends, perhaps related to loss of shoreline sources of sediment. Distinguishing between these possibilities is a topic of future research.

3.2. Sedimentological changes induced by breakwater construction

Observations at two breakwaters exemplify processes occurring at young (Elk Neck) and older (Eastern Neck) structures. Elk Neck, constructed in 2005, is an offshore semi-circular segmented breakwater. Both cores at Elk Neck display steady-state sedimentation and an accumulation rate of ~1 cm/y (Fig. 4). This is expected, because this breakwater was constructed only 3 years prior to sampling – a short time relative to the half-life of ^{210}Pb (22.3 y). Clues as to how much and what type of sediment deposited post-construction can be found in the grain-size data. Median diameter decreases (increasing phi units) from 1.5 to 3.0 phi (363.5 to 129.4 μm) at ~80 cm behind the breakwater, and from 1.5 to 3.1 phi (361.0 to 120.7 μm) at ~60 cm adjacent to it. Also, sediment fines upward behind the breakwater at ~20 cm, which is the expected trend in response to reduced physical energy; this trend is absent adjacent to the breakwater. In particular, median diameters observed in cores behind and adjacent to the breakwater deviate at 14 cm, yielding a post-construction sedimentation rate of 4.6 cm/y. Thus, sediment is significantly finer and more organic (surficial value of 2.1% compared to an average of 0.2% for the rest of the core) behind the breakwater. This trend of increasing fine and organic material may ultimately prove detrimental to SAV, outweighing the apparent initial benefit of the breakwater in alleviating potential wave limitation [21; Fig. 1].

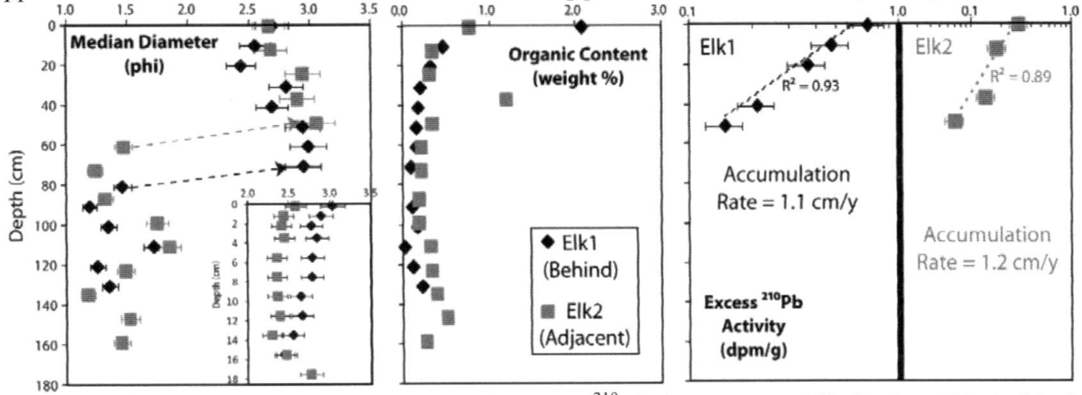

Figure 4. Grain size, organic content, and excess ^{210}Pb for cores at Elk Neck. Elk 1 (black diamonds) is landward of the breakwater; Elk 2 (grey squares) is adjacent to it at a similar water depth. Note that x-axes vary among profiles. Arrows in the right panel denote a shift toward finer sediment, occurring at ~80 and ~60 cm for Elk 1 and 2, respectively. Right panel, inset. Median diameters observed in surface sediments at Elk 1 and 2.

The record at Eastern Neck, sampled 15 years post-construction, shows different trends (Fig. 4). Grain size is similar in the lower portion of both cores, with a relatively coarse base below 177 cm – average median diameter 3.1 phi (119.8 μm) – overlain by sediment that fines upward to an average median diameter of 7.1 phi (7.1 μm) at 65 cm. Above this depth, sediment behind the breakwater coarsens, whereas it remains relatively unchanged adjacent to the breakwater. Adjacent to the breakwater, steady-state sedimentation dominates at a rate of 2.8 cm/y. However, there is an apparent change in accumulation rate behind the breakwater at 108 cm. The lower portion shows steady-state sedimentation and a rate of 2.6 cm/y, similar to the adjacent core. The upper portion has more variable activities, indicating episodic sedimentation, and an accumulation rate of 4.9 cm/y, though only 21% of the variation in excess ^{210}Pb is explained by radioactive decay with depth. At the surface of the core is a low-activity ~36-cm sand layer that is not included in the accumulation rate calculations. This sand layer likely corresponds to that appearing between the 2005 and 2007 aerial images (see Fig. 1). If this sand layer is assumed to be a recent event layer with respect to ^{210}Pb, and the change in accumulation rate at 108 cm is assumed to coincide with breakwater construction, then 72 cm of sediment accumulated within the last 15 y, yielding a rate of 4.8 cm/y.

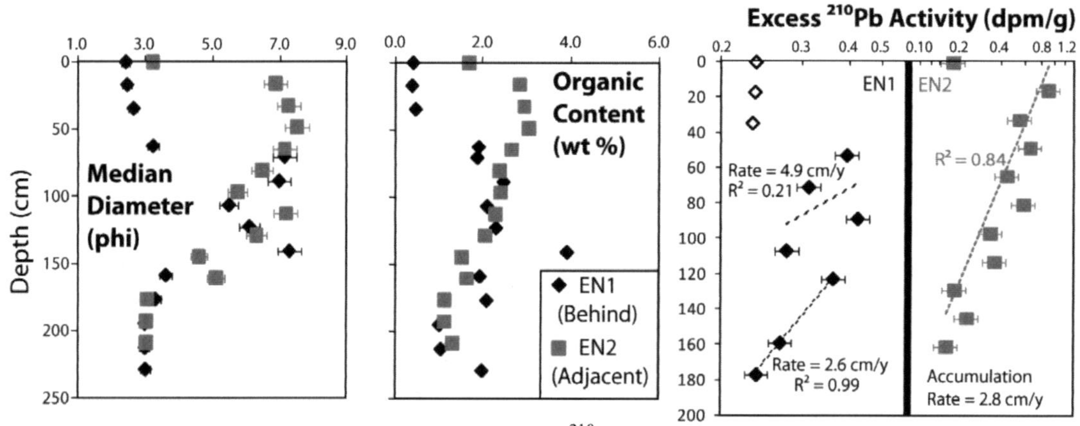

Figure 5. Grain size, organic content, and excess ^{210}Pb for cores at Eastern Neck. EN 1 (black diamonds) is landward of the breakwater; EN 2 (grey squares) is adjacent to it at a similar water depth. Note that x-axies vary among profiles. Open diamonds in the left panel denote total ^{210}Pb activity; these activities are not included in the regression fit.

While both locations show the expected increase in post-construction sedimentation rate behind the breakwater, the character of sediment deposited post-construction differs. This is likely due to the difference in sediment source. The Susquehanna River dominates sediment delivery in northern Chesapeake Bay, whereas mid-Bay locations are mostly influenced by shoreline sources [13]. Thus, fine material is available to accumulate behind the breakwater at Elk Neck once the physical energy is decreased. At Eastern Neck, both core locations likely respond to fining of shoreline-derived sediments similar to that previously observed in mainstem Chesapeake Bay [7] and in Crab Alley Bay. The largest difference between the two cores is a thick sand layer behind the breakwater, which may have been deposited by natural (longshore transport) and/or anthropogenic (sand replenishment) means. The response of SAV to these sedimentological changes is a topic of current study, as is extending the data set to include breakwaters of varying ages throughout the Chesapeake Bay.

4. Concluding Remarks

Sediment characteristics in nearshore Chesapeake Bay depend on the local sedimentary environments. In some locations (Crab Alley Bay, Eastern Neck), sediments have become finer and more organic over time, probably in response to land-use changes that have increased the supply of fine material. However, others have become less organic and coarser over time (Trippe Bay), which may relate to variations in the sedimentary environment induced by shoreline changes. Knowledge of these longer-term trends is essential for placing any potential changes due to recent shoreline modifications (e.g., hardening in Crab Alley Bay) in the proper context. More recently, construction of breakwaters has altered the sedimentary environment, changing the character of sediment deposited post-construction, with a 2-4-fold increase in sediment accumulation in the protected area. This sediment can be high in fine and organic content, when supply is available (e.g., Elk Neck), or it can be low-organic coarse sediment from longshore transport and/or material placement (e.g., Eastern Neck). Thus, because sedimentary processes vary according to physical setting, local trends must be discerned to determine whether a given site may be suitable for SAV restoration.

Acknowledgements
The authors thank NOAA Chesapeake Bay Office and the US Army Corps of Engineers for funding portions of the work presented here. We also thank Dale Booth, Caroline Baumgartner, and Taylor Robinson for invaluable field and laboratory assistance. The comments of three anonymous reviewers are gratefully acknowledged. This is UMCES contribution 4388.

References

[1] Allen JR 1988 Nearshore sediment transport *Geogr. Rev.* **78** 148-157
[2] Arnold RR, Cornwell JC, Dennison WC and Stevenson JC 2000 Sediment-based reconstruction of submersed aquatic vegetation distribution in the Severn River, a sub-estuary of Chesapeake Bay *J Coastal Res* **16** 188-195
[3] Batiuk RA, Orth RJ, Moore KA, Dennison WC and Stevenson JC 1992 Chesapeake Bay submerged aquatic vegetation habitat requirements and restoration targets: a technical synthesis *Virginia Institute of Marine Science report CBP/TRS-83/92*
[4] Berman MR, Berquist H, Dewing S, Hershner CH, Rudnicky T, Barbosa A, Woods H, Schatt DE, Weiss D and Rea H 2003 Dorchester County, Maryland Shoreline Situation Report *Comprehensive Coastal Inventory Program, Virginia Institute of Marine Science*
[5] Brush GS 2009 Historical land use, nitrogen, and coastal eutrophication: a paleoecological perspective *Estuaries Coasts* **32** 18-28
[6] Brush GS and Hilgartner WB 2000 Paleoecology of submerged macrophytes in the upper Chesapeake Bay *Ecol. Monogr.* **70** 645-667
[7] Cooper SR and Brush GS 1991 Long-term history of Chesapeake Bay anoxia *Science* **245** 992-996
[8] deBoer WF 2007 Seagrass-sediment interactions, positive feedbacks and critical thresholds for occurrence: a review *Hydrobiologia* **591** 5-24
[9] Dennison WC, Orth RJ, Moore KA, Stevenson JC, Carter V, Kollar S, Bergstrom PW and Batiuk RA 1993 Assessing water-quality with submersed aquatic vegetation *Bioscience* **43** 86-94
[10] Erftemeijer PLA and Koch EW 2001 Sediment geology methods for seagrass habitat *Global Seagrass Research Methods* ed Short FT and Coles RG (Elsevier Science, London) pp 345-367
[11] Hardaway CS and Byrne RJ 1999 Shoreline Management in Chesapeake Bay *Virginia Institute of Marine Science, Report VSG-99-11*
[12] Heck KL, Able KW, Roman CT and Fahay MP 1995 Composition, abundance, biomass, and production of macrofauna in a New-England estuary – comparisons among eelgrass meadows and other nursery habitats *Estuaries* **18** 379-389
[13] Hobbs CH, III, Halka JP, Kerhin RT and Carron MJ 1992 Chesapeake Bay sediment budget *J. Coastal Res.* **8** 292-300
[14] Kemp WM, Boynton WR, Adolf JE, Boesch DF, Boicourt WC, Brush G, Cornwell JC, Glibert PM, Hagy JD, Harding LW, Houde ED, Kimmel DG, Miller WD, Newell RIE, Roman MR, Smith EM and Stevenson JC 2005 Eutrophication of Chesapeake Bay: historical trends and ecological interactions *Mar. Ecol. Progr.* **303** 1-29
[15] Kennedy VS 1996 The ecological role of the eastern oyster, Crassostrea virginica, with remarks on disease *J. Shellfish Res.* **15** 177-183
[16] Koch EW, Wazniak C, Karrh L and Wells D in prep Sediment as a limiting factor to seagrass

distribution in the Coastal Bays of Maryland, USA

[17] Koch EW 2001 Beyond light: physical, geological, and geochemical parameters as possible submersed aquatic vegetation habitat requirements *Estuaries* **24** 1-17

[18] Landwehr JM, Reel JT, Rybicki NB, Ruhl HA and Carter V 1999 Chesapeake Bay habitat criteria scores and the distribution of submersed aquatic vegetation in the tidal Potomac River and estuary *USGS Open File Report 99-219*

[19] Livingston RJ, McGlynn SE and Xufeng N 1998 Factors controlling seagrass growth in a gulf coastal system: water and sediment quality and light *Aquat. Bot.* **60** 135-159

[20] Lubbers L, Boynton WR and Kemp WM 1990 Variations in structure of estuarine fish communities in relation to abundance of submersed vascular plants *Mar. Ecol. Progr.* **65** 1–14

[21] Madsen JD, Chambers PA, James WF, Koch EW and Westlake DF 2001 The interaction between water movement, sediment dynamics, and submersed macrophytes *Hydrobiologia* **444** 71-84

[22] Martin D, Bertasi F, Colangelo MA, de Vries M, Frost M, Hawkins SJ, Machpherson E, Moschella PS, Satta MP, Thompson RC and Ceccherelli VU 2005 Ecological impacts of coastal defense structures on sediment and mobile fauna: evaluating and forecasting consequences of unavoidable modifications of native habitats *Coast. Eng.* **52** 1027-1051

[23] Maryland Geological Survey 2000a Historical shorelines, 1841-1976; Chesapeake Bay region of Maryland Baltimore, Maryland, USA

[24] Maryland Geological Survey 2000b Recent (ca. 1990) shorelines, coastal regions of Maryland Baltimore, Maryland, USA

[25] Meadows GA, Mackey SD, Goforth RR, Mickelson DM, Edil TB, Fuller J, Guy DE, Meadows LA and Brown E 2005 Cumulative habitat impacts of nearshore engineering *J. Great Lakes Res.* **31** 90-112

[26] Moore KA, Neckles HA and Orth RJ 1996 *Zostera marina* (eelgrass) growth and survival along a gradient of nutrients and turbidity in the lower Chesapeake Bay *Mar. Ecol. Progr.* **142** 247–259

[27] Orth RJ and Moore KA 1983 Chesapeake Bay: an unprecedented decline in submerged aquatic vegetation *Science* **222** 51–53

[28] Orth RJ and Moore KA 1984 Distribution and abundance of submerged aquatic vegetation in Chesapeake Bay - an historical perspective *Estuaries* **7** 531-540

[29] Orth RJ, Wilcox DJ, Whiting JR, Nagey LS, Owens AL and Kenne AK 2008 2007 distribution of submerged aquatic vegetation in Chesapeake Bay and Coastal Bays *Virginia Institute of Marine Science Special Scientific Report 151*

[30] Palinkas CM and Nittrouer CA 2006 Clinoform sedimentation along the Apennine shelf, Adriatic Sea *Mar. Geol.* **234** 245-260

[31] Palinkas CM and Koch EW 2009 SAV and its substrate: spatial, temporal, and historical trends. *Final Report to NOAA Chesapeake Bay Office*

[32] Rabalais NN, Turner RE, Sen Gupta BK, Platon E and Parsons ML 2007 Sediments tell the hisory of eutrophication and hypoxia in the northern Gulf of Mexico *Ecol. Appl.* **17** S129-S143

[33] Ranasinghe JA, Weisberg SB, Smith RW, Montagne DE, Thompson B, Oakden JM, Huff DD, Cadien DB, Velarde RG and Ritter KJ 2009 Calibration and evaluation of five indicators of benthic community condition in two California bay and estuary habitats *Mar. Pollut. Bull.* **59** 5-13

[34] Stone GW and Stapor FW, Jr 1996 A nearshore sediment transport model for the northeast Gulf of Mexico *J. Coastal Res.* **12** 786-793

[35] Willams M, Longstaff B, Buchanan C, Llanso R and Dennison W 2009 Development and evaluation of a spatially-expicit index of Chesapeake Bay health *Mar. Poll. Bull.* **59** 14-25

IOP Conf. Series: Earth and Environmental Science **9** (2010) 012013 doi:10.1088/1755-1315/9/1/012013

Temporal trend in the intensity of subsurface saltwater ingressions to coastal Lake Sarbsko (northern Poland) during the last few decades

M Woszczyk[1], W Spychalski[2], M Lutyńska[1] and R Cieśliński[3]
[1] Department of Quaternary Geology and Paleogeography, Adam Mickiewicz University, Dzięgielowa 27, 61-680 Poznań, Poland
[2] Department of Soil Science, Poznań Uniwersity of Life Sciences, Szydłowska 50, 60-656 Poznań, Poland
[3] Department of Hydrology, University of Gdańsk, Dmowskiego 16a, 80-952 Gdańsk, Poland

E-mail: woszczyk@amu.edu.pl

Abstract. The present paper provides evidence of the increasing intensity of subsurface saltwater ingressions (SSI) to coastal Lake Sarbsko (northern Poland) during the last few decades. Saltwater ingressions to the lake were recognized by spatial and seasonal changes in the concentrations of chlorides in pore waters from the lake surface sediments. Temporal trends in SSI were reconstructed based on diatom proxies from a lake sediment core located within the range of saltwater ingressions. The data showed that subsurface saltwater input to Lake Sarbsko occurs during severe winter storms on the Baltic Sea, when saltwater is pumped to the lake from under the barrier separating the lake from the sea. Due to accelerated sea level rise, increased frequency of storms and lack of ice cover on the Baltic since at least the mid 20th century, saline groundwater supply to the lake has distinctly increased.

1. Introduction

Accelerated sea level rise is one of the most discernible effects of the global environmental changes occurring over the last few decades. Sea level rise causes inundation of low-lying coastal plains and deltas, increases the probability of storm surges as well as intensifies coastal erosion and salinization of freshwater coastal basins and groundwater aquifers [1] – [8].

Subsurface saltwater ingressions to freshwater aquifers were shown to have considerable economical, ecological and geochemical impacts. For example, deterioration of potable water quality in densely populated coastal areas was reported by [9] – [14]. The influence of submarine groundwater discharges on the trophic status and bioproductivity in the fresh- and saltwater mixing zones was shown by [15] – [18] and secondary geochemical changes in sediments (i.e. due to cation exchange phenomena, dissolution of feldspars, carbonates, clay minerals etc.) caused by percolating saltwaters were postulated by [8] [16] [18] [19] [20].

SSI intensity and its temporal changes on the coast of the southern Baltic have not been monitored so far. The present paper is the first attempt to demonstrate increasing intensity of saline groundwater ingressions to coastal lakes (closed lagoons) in the last few decades using Lake Sarbsko as an example.

Lake Sarbsko is situated on the Gardno-Łeba Coastal Plain (northern Poland) (17°27'E, 54°46'N, 0.5 m asl) and is a shallow freshwater coastal lake. It covers 6.5 km^2 and its average depth is 1.5 m with a maximum of 3.5 m (figure 1). The main surface inflow is the Chełst River from the east and a small dyke from the south. The outflow is situated in the western part of the basin (the Chełst outflow channel, Młynówka channel) and connects Lake Sarbsko with the Łeba River mouth (figure 1).

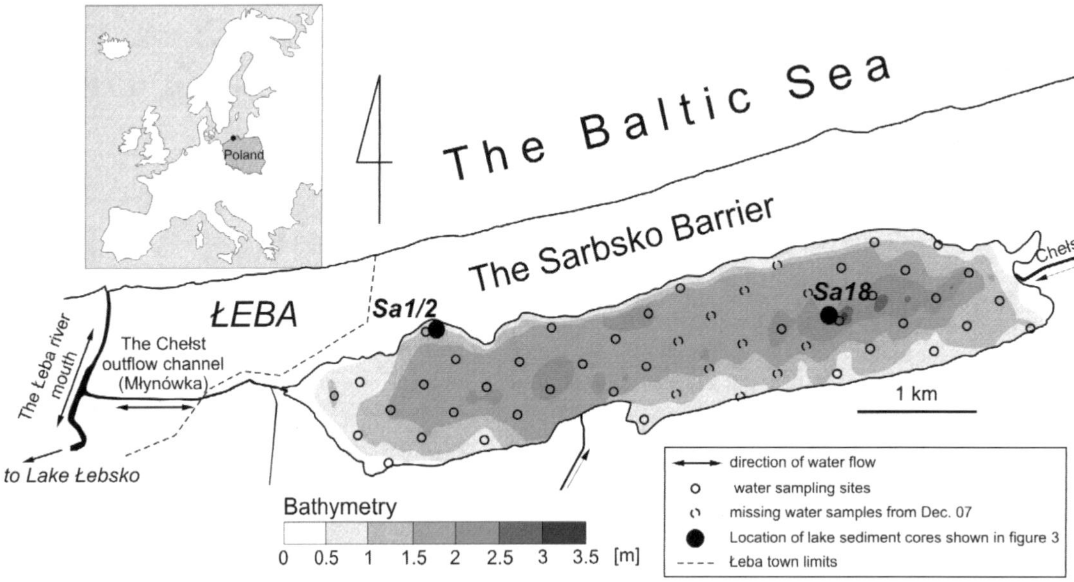

Figure 1 Location of a study area, distribution of pore water sampling points and coring sites.

In the paper, SSI to Lake Sarbsko is demonstrated by temporal changes in pore water chlorinity as Cl^- ions usually behave conservatively during estuarine mixing [21]. Moreover, chlorides are not adsorbed on sediment particles and not subject to early diagenetic processes [22]. Chlorinity was earlier used to delimit spatial extent of salt water ingressions in coastal aquifers [12] [20]. Changes in the intensity of saltwater ingressions in the past are showed by paleoecological data from a lake sediment core located within the range of saltwater ingressions in the Lake Sarbsko. To show that temporal trends in salinity in the lake are exclusively due to increasing SSI intensity, we compared the paleosalinity record from this first core with paleoecological data from another core located in a part of the lake not affected by saltwater.

2. Methods

Pore waters were sampled at 39 points throughout the lake on the 13th of December 2007 and at 50 stations on the 10th of April, 10th of July, 17th of September 2008 (figure 1). 5-cm thick samples of surface deposits together with interstitial waters were taken using gravity corer. Pore waters were extracted from the sediment by centrifugation and analyzed with respect to chlorinity. Chlorides were determined by titration with $AgNO_3$ using K_2CrO_4 as indicator. Analytical precision was better than 5%.

Sediment profiles Sa1/2 and Sa18 (figure 1, figure 3) were selected from a set of 45 cores recovered previously from Lake Sarbsko [23]. The main criterion of selection was location of the coring site relative to spatial changes in pore water chlorinity in the lake shown in figure 2. The profile Sa1/2 is situated within the range of SSI and Sa18 is located out of this zone.

The cores Sa1/2 and Sa18 were subsampled at 5 cm intervals, however in the latter diatoms were analyzed with 10-20 cm resolution. Description of chemical methods applied in the analysis of sediments can be found in [24]. Diatom analysis was carried out according to standard protocols [25] and diatoms were grouped by salinity preferences [26].

Radiocarbon ages from the Sa1/2 profile are after [24] and [27]. Calibration was made using OxCal 4.1 [28] (http://c14.arch.ox.ac.uk). Reservoir correction was not applied for the ^{14}C ages because the hard water effect in the modern Lake Sarbsko (which could be used for the past) seems to be considerably low (figure 3). In the NE section of the lake, sediment accumulation rate calculated based on ^{210}Pb activity profiles was 2-3 mm·y^{-1} [29]. If these results are adopted for the Sa1/2 coring site the theoretical age of a sample from 0.9 m below the lake bottom is c.a. 1550-1700 AD, which is even older than the lower limit of calibrated ^{14}C age from that level (figure 3).

3. Evidence of subsurface saltwater ingression to Lake Sarbsko

Despite the fact that Lake Sarbsko is a freshwater lake [30] isolated from the Baltic Sea by a continuous barrier (figure 1), the sea affects chemical composition of its waters by occasional surface salt water ingressions. As a rule, salt waters enter the lake during severe storms via the mouth of the Łeba River and the Chełst outflow channel [31]. Nevertheless, this effect does not account for spatial and temporal changes in the concentration of chlorides in pore waters of Lake Sarbsko (figure 2).

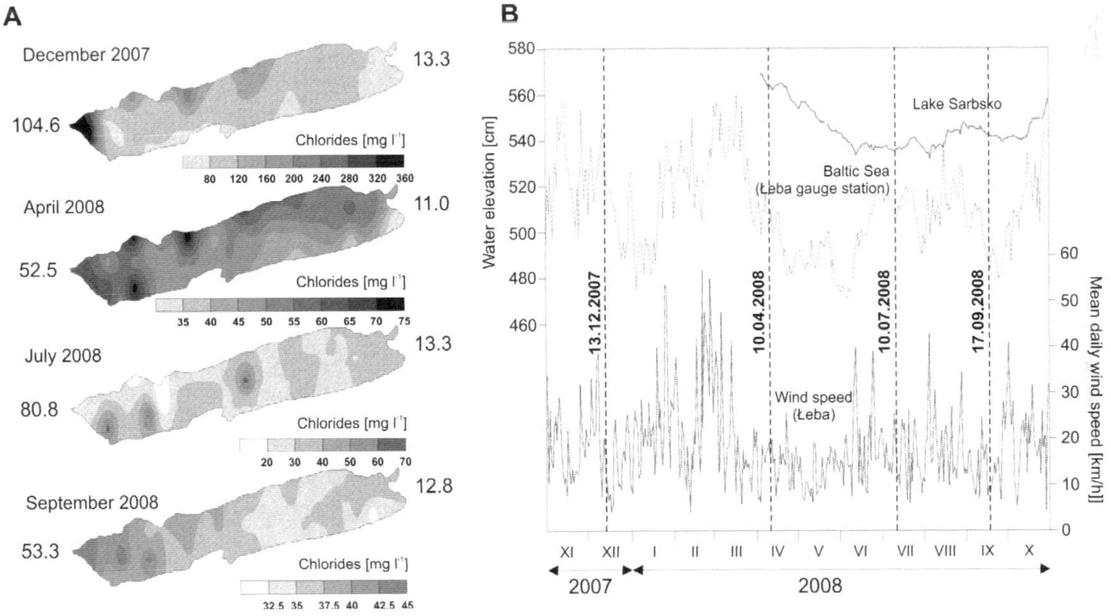

Figure 2. Chlorinity of pore waters from Lake Sarbsko (A) and water level in the lake and the Baltic Sea as well as daily mean wind speed in Łeba between the 1 November 2007 and 31 October 2008 (B). Numbers to the left and right of the maps in panel A indicate concentrations of Cl^- [mg l^{-1}] in the Młynówka channel and the Chełst River, respectively. Storm activities in the Baltic are indicated by increased water level and high wind speed. The lake water level record is available since 3 April 2008. Sea level data were obtained from the Institute of Meteorology and Water Management (IMGW) in Gdynia, Poland. Wind speed data are taken from www.tutiempo.net/en/Climate/LEBA/121200.htm. Vertical dashed lines indicate water sample collection dates in the notation dd.mm.yyyy.

In December 2007, Cl^- concentrations in pore waters in the lake varied between 38.8 mg l^{-1} and 351.7 mg l^{-1} with highest values recorded in westernmost part of the lake and just in front of the Sarbsko Barrier head (figure 2A). Interestingly, on December 2007, chloride concentrations in pore waters were generally higher than in bottom and surface waters as well as in the Młynówka channel (figure 2A). Similar patterns of Cl^- distribution in interstitial solutions, albeit with significantly lower

Cl^- concentrations (34.4 – 74.8 mg l^{-1}), were obtained in April 2008. On the contrary, in July 2008 pore waters throughout the lake were highly depleted in Cl^- (8.0 – 68.0 mg l^{-1}) and displayed no clear trend in spatial variation of chlorides. In September 2008, Cl^- concentrations in Lake Sarbsko water slightly increased compared to summer months (31.6 – 45.1 mg l^{-1}) and the gradients between western/northern and central parts of the basin started to build up again.

Spatial and temporal changes in pore water chlorinity seem to indicate that marine water percolates to Lake Sarbsko through the Sarbsko Barrier. If surface saltwater inflows via the Łeba River mouth and the Młynówka channel were the only reason for seasonal increases in pore water chlorinity in Lake Sarbsko, Cl^- concentrations in the lake waters should gradually diminish towards the east. Observed distribution of chlorides in the lake, with local isolated enrichments in Cl^- located along the barrier and at considerable distances from the Młynówka channel, does not fit this simple model. Moreover, maximum Cl^- concentrations in Lake Sarbsko coincide with increased Baltic Sea levels as well as enhanced storm activity on the southern Baltic from winter to early spring (figure 2B).

The mechanism of saltwater intrusions to Lake Sarbsko is not fully understood so far; however, it is anticipated that seawater is pumped into the lake during storm surges when the gradient in hydrostatic pressure develops between seaward and landward sides of the beach barrier [16] [18] [32].

4. Sedimentary record of temporal changes in intensity of saltwater ingressions to the lake

Sediments of Lake Sarbsko reveal considerable variations in the abundance of brackish water diatoms (figure 3) indicating temporal changes in salinity in this coastal lake.

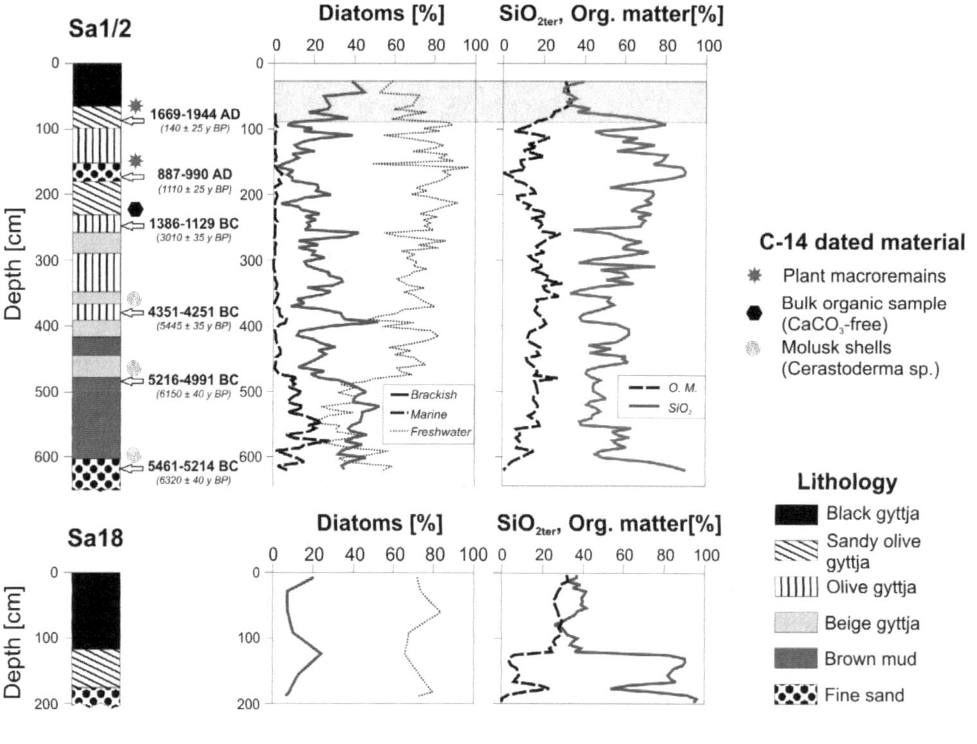

Figure 3. Chemical composition of sediments and diatom succession in the Sa1/2 and Sa18 cores. Location of coring sites given in figure 1. Increase in the content of brackish water diatoms in the top of the Sa1/2 core (marked in grey) is interpreted as the result of increased frequency of saltwater ingressions to Lake Sarbsko. The uppermost 15-cm layer of the Sa1/2 core is lacking due to fluffy texture of the sediments. The Sa18 core has not been dated.

Salinity of Lake Sarbsko waters varied due to both changes in the degree of surface connection between the back-barrier basin and the sea as well as subsurface hydraulic contact between the lake and marine waters.

Prior to stabilization of the Sarbsko Barrier close to its present location around 3500 – 3100 BC [33], [34], the sand barrier beach was frequently breached by severe storm surges which resulted in opening the straits between the lagoon and the sea. Consequently, marine water could easily entered the back-barrier basin thus increasing its salinity. Marine ingressions to the lagoon corresponded with the Baltic Sea level highstands [27] and sediment is marked with brown mud layers enriched in brackish water and marine diatoms (figure 3).

Between c.a. 660 – 1200 AD, when the lake was already isolated from the sea, short-term salinity increases in the lake can be attributed to overwash events. The latter are recorded in 0.3-m thick find sand and sandy olive gyttja (between 2.2 – 1.9 m in the Sa1/2 profile) with elevated contributions of brackish and redeposited marine diatoms (figure 3).

The top part of the sedimentary sequence of Lake Sarbsko is dominated by black gyttja (lake mud stained with FeS), which is of the main interest in the present study. Deposition of the facies commenced between ~1669 and ~1944 AD (figure 3). This imprecise radiocarbon date could not be improved by the ^{210}Pb method due to irregular unsupported lead activity curve in a short core recovered close to the Sa1/2 site [29]. The ^{210}Pb profile suggest the possibility of resuspension of the sediments [29], however, distinct gradients in bulk geochemical parameters as well as in the abundance of brackish and freshwater diatoms in the top part of the Sa1/2 core (figure 3) argue against substantial mixing [35].

Relatively low detrital (terrigenous) silica content, increased organic matter concentration and the lack of marine diatoms in the black gyttja (figure 3) seem to indicate that the facies was deposited in a closed lake sheltered from marine ingressions by the well developed barrier. Nevertheless, increased abundances of brackish diatoms in the black gyttja (up to 40%) reflect increased salinity of lake water.

Increased salinity of Lake Sarbsko water in the last decades seems to be of local significance, as it has been recognized neither in the central-eastern part of Lake Sarbsko (the profile Sa18 does not show any significant increase in brackish diatoms in the top part of the core – figure 3) nor in other lakes situated in the southern Baltic coastal zone [36] [37] [38]. These observations support the idea developed in the previous section that increased salinity in the northern part of Lake Sarbsko, recorded in the upper section of the Sa1/2 profile, resulted from percolation of salt waters from under the Sarbsko Barrier.

Moreover, increase in the abundance of brackish diatom species within uppermost section of the profile may indicate that SSI intensity in Lake Sarbsko has increased since at least mid-20th century. This increase can be linked on one hand to the Baltic Sea level rise and, on the other hand, to intensive exploitation of groundwater in the vicinity of Łeba. The second alternative seems less important because Holocene-Pleistocene aquifer, which is in direct hydraulic contact with Lake Sarbsko waters, is only used to a limited extent for economical purposes [39]. Drinking water in Łeba is mainly exploited from a lower Pleistocene aquifer separated from the upper Pleistocene and Holocene-Pleistocene ones by impermeable glacial tills [39]. As a matter of fact, in the 1980s in the Holocene-Pleistocene aquifer in the vicinity of Łeba local short-term chlorinity variations were observed. However, an increasing tendency was not recorded and higher salinity of groundwater was attributed to subsurface ingression of Lake Łebsko waters [39].

The Baltic Sea level has risen since at least the mid 19th century [40]. Between 1951 and 1990 the magnitude of this process on the middle Polish coast (in Ustka, 30 km westwards from Lake Sarbsko) was 4.38 mm·y^{-1} [41]. The rise in mean sea level was accompanied by the increase in the frequency of storm surges, which are one of the main driving forces for SSI. Over the same time period the frequency of high sea levels in Ustka increased by 101% [41]. The permanent lack of ice cover on the southern Baltic in recent years is an additional factor for seawater percolation into Lake Sarbsko. The absence of sea ice enables a direct subsurface hydraulic connection between the lake and marine waters during storm seasons. Since 1965 the Baltic Sea in the vicinity of Łeba has been iced over for only 164 days (data obtained in the Institute of Meteorology & Water Management, Gdynia, Poland).

5. Conclusions

Hydrochemical analyses of Lake Sarbsko waters revealed that, during severe storm surges on the Baltic Sea, the lake receives subsurface input of saltwaters from under the Sarbsko Barrier. This results in local increases in salinity in the basin.

Due to accelerated sea level rise in the last few decades, the intensity of saline groundwater supply to the lake has distinctly increased. The onset of this process is not known so far due to the lack of precise chronology in the youngest sediments of Lake Sarbsko. However, ongoing research on supplementary sediment cores from the lake are expected to fill this gap in the next few months.

Acknowledgments

The study was made possible due to financial support of the Polish Ministry of Science and Higher Education (grant no. NN306 3543 33). John Smol and Adi Torfstein are acknowledged for their valuable comments and suggestions that helped to improve this paper.

References

[1] Titus J G 1990 *Changing Climate and the Coast* vol 1, ed J G Titus et al (Washington: US Army Corps of Engineers, Environmental Protection Agency, National Oceanic and Atmospheric Administration) 63-85
[2] Awosika L F, Ibe A C, Udo-Aka M A 1990 *Changing Climate and the Coast* vol 2, ed J G Titus et al (Washington: US Army Corps of Engineers, Environmental Protection Agency, National Oceanic and Atmospheric Administration) 49-65
[3] Sestini G 1990 *Changing Climate and the Coast* vol 2, ed J G Titus et al (Washington: US Army Corps of Engineers, Environmental Protection Agency, National Oceanic and Atmospheric Administration) 115-125
[4] Jelgersma S 1990 *Changing Climate and the Coast* vol 2, ed J G Titus et al (Washington: US Army Corps of Engineers, Environmental Protection Agency, National Oceanic and Atmospheric Administration) 241-245
[5] Moreira M E S A 1990 *Changing Climate and the Coast* vol 2, ed J G Titus et al (Washington: US Army Corps of Engineers, Environmental Protection Agency, National Oceanic and Atmospheric Administration) 293-308
[6] Muehe D, Neves C F 1990 *Changing Climate and the Coast* vol 2, ed J G Titus et al (Washington: US Army Corps of Engineers, Environmental Protection Agency, National Oceanic and Atmospheric Administration) 311-327
[7] Rotnicki K, Borówka R K, Devine N 1995 *J. Coast. Res.* **Sp. issue 22** 111-35
[8] Cardona A, Carrillo-Rivera J J, Huizar-Álvarez R, Graniel-Castro E 2004 *Env. Geol.* **45** 350-66
[9] Melloul A J, Goldenberg L C 1997 *J. Env. Manag.* **51** 73-86
[10] Pulido-Bosch A, Tahiri A, Vallejos A 1999 *Wat. Air Sci. Pol.* **114** 323-37
[11] Marui A 2003 *Geol. Quat.* **47(4)** 381-88
[12] Demirel Z 2004 *J. Env. Manag.* **70** 275-282
[13] Park S C, Yun S T, Chae G T T, Yoo I S, Shin K S, Heo C H, Lee S K 2005 *J. Hydrol.* **313** 182-94
[14] Antonellini M, Mollema P, Giambastiani B, Bishop K, Caruso L, Minchio A, Pellegrini L, Sabia K, Ulazzi E Gabbianelli G 2008 *Hydrogeol. J.* **16** 1541-56
[15] Osgood D T 2000 *Wet. Ecol Manag.* **8** 133-46
[16] Burnett W C, Bokuniewicz H, Huettel M, Moore W S, Taniguchi M 2003 *Biogeochemistry* **66** 3-33

[17] Taniguchi M, Burnett W C, Smith C F, Paulsen R J, O'Rurkie D, Krupa S L, Christoff J L 2003 *Biogeochemistry* **66** 35-53
[18] Michael H A, Mulligan A E, Harley Ch F 2005 *Nature* **46** 1145-48
[19] Burzyński K, Kozerski B, Sadurski A 1999 *Biuletyn PIG* **388** 35-48
[20] Petalas C, Lambrakis N 2006 *J. Hydrol.* **324** 51-64
[21] Liss P S 1976 *Estuarine Chemistry*, ed J D Burton and P S Liss (London, New York, San Francisco: Academic Press) 93-130
[22] Berner R A 1980 *Early Diagenesis A Theoretical Approach* (Princeton: Princeton University Press) 241pp
[23] Woszczyk M 2004 *Geneza, litologia i stratygrafia osadów czwartorzędowych* vol. 4, ed A Kostrzewski (Poznań, UAM) 487-95
[24] Bechtel A, Woszczyk M, Reischenbacher D, Sachsenhoffer R, Gratzer R, Püttmann W, Spychalski W 2007 *Org. Geoch.* **38** 1112-31
[25] Battarbee R W 1986 *Handobook of Holocene Palaeoecology and Palaeohydroogy*, ed B E Berglund (London: Wiley & Sons Ltd.) 527-70
[26] Denys L 1991/92 *Serv. Geol. de Belgique* 1,2
[27] Woszczyk M, Lutyńska M, Spychalski W 2008 *Holoceńskie przemiany wybrzeży i wód południowego Bałtyku – przyczyny, uwarunkowania i skutki*, ed K Rotnicki et al (Poznań-Bydgoszcz: Wyd. Znak) 145-54
[28] Bronk Ramsey C 1995 *Radiocarbon* **37(2)** 425-30
[29] Szarafin T 2009 *Współczesna sedymentacja osadów w jeziorze Sarbsko* (Gdynia: UG) unp. M.Sc Thesis 45pp
[30] Cieśliński R 2004 *Lim. Rev.* **4** 57-66
[31] Szopowski Z 1962 *Materiały do monografii polskiego brzegu morskiego*, ed Z Szopowski (Gdańsk-Poznań: PWN) 137-56
[32] Rapaglia J, Di Sipio E, Bokuniewicz H, Zuppi G M, Zaggia L, Galgaro A, Beck A 2010 *Cont. Shelf Res.* **30** 119-126
[33] Rotnicki K, Borówka R K, Pazdur A, Hałas S, Krzymińska J 1999 *Quat. Stud. Pol.* **16** 67-79
[34] Tobolski K 1979 *Bad. Fizjogr. n. Pol. Zach.***32 Seria A** 151-68
[35] Oldfield F, Appleby P G 1984 *Lake Sediment and Environmental History*, ed E Y Haworth and J W G Lund (Leicester: Leicester Univ. Press) 93-124
[36] Przybyłowska-Lange W 1979 *Acta Palaeobot.* **20 (2)** 227-44
[37] Wojciechowski A 1987 *Bad. Fizjogr. n. Pol. Zach.* **37 Seria A** 191-209
[38] Wojciechowski A 1995 *Jour. of Coast. Res.* **Sp. issue 22** 237-43
[39] Kwaterkiewicz A, Sadurski A, Zuber A 1999 *Współczesne problemy hydrogeologii* vol IX 187-94
[40] Zeidler R B, Wróblewski A, Miętus M, Dziadziuszko Z, Cyberski J 1995 *Jour. Coast. Res.* **Sp. issue 22**, 33-55
[41] Rotnicki K, Borzyszkowska W 1999 *Ewolucja geosystemów nadmorskich południowego Bałtyku*, ed R K Borówka et al (Poznań-Szczecin, Bogucki Wyd. Nauk) 141-60

IOP Conf. Series: Earth and Environmental Science 9 (2010) 012014 doi:10.1088/1755-1315/9/1/012014

Mid-late Holocene environments of Agua Buena locality (34°50′S; 69°56′W), Mendoza, Argentina

D Navarro[1,2]*, A Mehl[3], M A Zarate[3] and M M Paez[1]

[1] Laboratorio de Paleoecología y Palinología, Departamento de Biología, Universidad Nacional de Mar del Plata, Funes 3350, (B7602AYL) Mar del Plata, Buenos Aires, Argentina.

[2] CONICET, Consejo Nacional de Investigaciones Científicas y Técnicas, Argentina.

[3] INCITAP - Instituto de Ciencias de la Tierra y Ambientales de La Pampa. CONICET- Universidad Nacional de La Pampa, Uruguay 151, (L6300CLB) Santa Rosa, La Pampa, Argentina.

*E-mail: dnavarro@mdp.edu.ar, navarro.dd@gmail.com

Abstract. In southern South America the acquisition of high-quality Holocene paleoclimate data is a priority due to the paucity of complete, continuous and well dated records. Here we report preliminary results from a combined sedimentological and palynological study of an alluvial fan sequence and the laterally connected sedimentary deposits of the Vega de la Cueva profile at Agua Buena east of the Andes in central Argentina. The main geomorphological units of the area were identified and mapped based on satellite image analysis and multiple field surveys. The sedimentological and pollen results allowed us to reconstruct the development of some environments. The Agua Buena record corresponds to the distal facies of the Arroyo Bayo alluvial fan starting the aggradation process prior to ca. 4100 cal yr BP. The organic-rich levels found were formed during the development of wetlands (vegas) dominated by Cyperaceae, Juncaceae and Poaceae. These highly productive environments with almost permanent water saturation were important between 4100 and 2800 cal yr BP, indicating more stable conditions. After 2800 cal yr BP, the organic content was comparatively lower with increasing sedimentation rates that are indicative of higher fluvial discharges. This information is fundamental to interpret both the pollen and charcoal records of the area and to evaluate their representativeness and potential to reconstruct past local and/or regional vegetation.

1. Introduction

In arid regions, water is the limiting factor for the development of vegetation, soils and agriculture. Consequently knowing past precipitation changes is essential to predict possible future environmental scenarios and their impact on the availability of natural resources. In recent years various interdisciplinary studies allowed the development of paleoclimatic models for southern South America and to propose regional and interhemispheric climatic linkages [1, 2, 3, 4, 5, 6, 7, among others]. Most of them were developed within the framework of the Pole-Equator-Pole (PEP I) initiative as part of the International Geosphere Biosphere Programme-Past Global Changes project (IGBP-PAGES). Nevertheless, vast areas of Argentina still lack continuous paleoecological records reaching back to the mid-Holocene and most of the literature refers to the need of complete, continuous and well-dated records. Furthermore, most of the existing paleoenvironmental studies focused on the late Pleistocene-Holocene transition, particularly the last glacial Maximum and postglacial intervals. Several scenarios have been proposed to explain the dynamics of the climate system during these intervals of dramatic environmental changes. However, only a scarce number of records are available for testing these hypotheses, particularly during the present interglacial. Therefore, the acquisition of high-quality Holocene paleoclimate data with a broad geographical distribution is a priority in order to postulate more realistic climatic scenarios.

In order to fill this gap we performed paleoecological studies during the last three years in the upper basin of the Atuel River (Andes) including the palynological and sedimentological analysis of an alluvial fan sequence (Agua Buena), and of a sediment core from Lake El Sosneado [8, 9] (Figures 1 and 2). The main goal is to reconstruct the environmental history during the mid-late Holocene and discuss the paleoclimatic significance of the records. In this contribution we report the preliminary results from the sedimentological and palynological analysis of a section outcropping at the locality of Agua Buena.

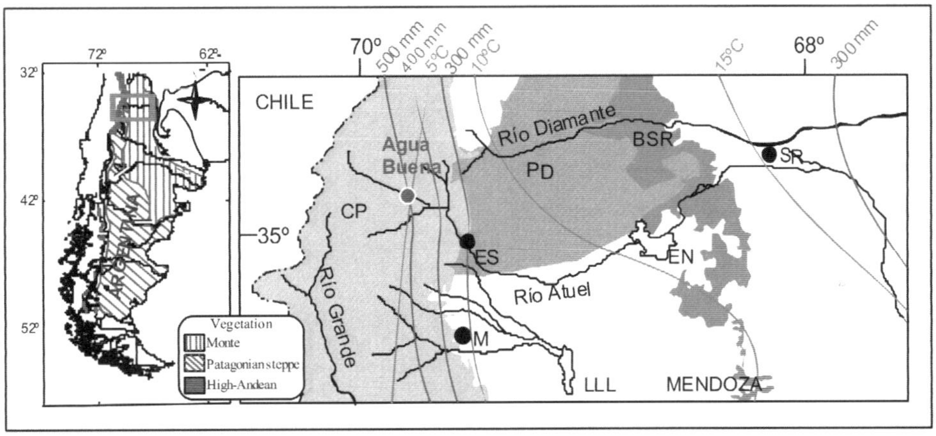

Figure 1. The map on the left displays the main vegetation units. The area shown by the orange rectangle is enlarged in the right panel, showing the location of the studied sites and the mean annual temperature (isotherms in orange) and precipitation (isohyets in blue) in the area. BSR indicates Bloque de san Rafael; CP Cordillera Principal; EN, El Nihuil; ES, El Sosneado; LLL, Laguna Llancanelo; M, Malargue; PD, Pampa del Diamante; SD, Salina del Diamante; SR, San Rafael.

1.1. Study Area

The Agua Buena locality is situated in the upper basin of the Atuel River (Figure 1; 34°55′-34°40′S and 69° 35′-70° 10′S; 2000-3000 m asl), which drains the Andes Cordillera. It is a key climatic location corresponding to the transition between the arid-semiarid conditions of the central Andes of Cuyo and the humid conditions of Patagonia to both the west and the south [10]. The Andes, with the highest peaks in the area ranging between 3000-4000 m asl acts as a geographical barrier for most of the humidity carried by the westerly winds coming from the Pacific Ocean. Nevertheless, the comparatively lower altitude of the Cordillera at this latitude allows part of the moist westerly winds to reach the eastern side of the Andes. As a result, the climate is semi-arid on the eastern Andean side with a winter precipitation regime (600 to 400 mm mean precipitation) extending from April to October, and a mean annual temperature of 5°C [11, 12]. Modern vegetation comprises two phytogeographic provinces [13]. The High-Andean province (>2400 m asl) is composed of an open grass steppe and cushion shrub steppe with occasional wetlands (locally known as "vegas"). The Patagonia province (2400-1600 m asl) is formed by shrub steppes, cushion shrub steppes, with grass steppes and wetlands in humid areas. Towards the east of the study area the Monte shrub steppes develop at altitudes of less than 1600 m asl.

1.2. Previous studies

At Agua Buena, a terminal moraine located upstream and superimposed on the late-Pleistocene terminal moraine of the Arroyo Malo Glaciation (2000 m asl) has been previously described [14, 15, 16]. The authors of these papers inferred a significant mid-Holocene neoglacial advance based on a radiocarbon age of 4080±75 yr BP by Stingl and Garleff [14], obtained from a peat-like deposit located at the base of what they interpreted as an alluvial terrace developed after the glacial retreat. Subsequently, these moraine deposits were reinterpreted as postglacial debris slides superimposed on the late-Pleistocene moraines [17]. Furthermore, the mid-Holocene terminal moraines were described 30 km upstream and 8 km away from the modern glacial front that is situated around 2800 m asl since 1948 AD [18, 19]. According to Gosse and Evenson [17] the radiocarbon age of 4080±75 yr BP [14] indicates a minimum age of the damming of the Atuel River by the debris slides. The latter formed a large lake of which today only remains the modern Lake El Sosneado as well as lake sediments outcropping in the valley sides.

2. Methods

The main geomorphological units of the area were identified and mapped based on satellite image analysis and multiple field surveys. A facies analyses approach was used to describe two stratigraphic sequences, Agua Buena (AB) and Vega de la Cueva (VC) [20, 21]. Lithological (color, texture, sedimentary structures, geometry) and pedological attributes (color, texture, pedological structure and features, roots, nodules, and consistency) were taken into account to identify the lithofacies. The boundaries of the sedimentary levels were described following [22].

Table 1. Uncalibrated radiocarbon dates and calibrated ages from Vega de la Cueva and Agua Buena sequences.

Site[a]	Uncalibrated ^{14}C age (^{14}C yr BP)[b]	Calibrated age (cal yr BP) with 2 sigma range[c]	Material dated	Lab. Reference[d]
VC5	427 ± 34	**459** (326-407, 436-507)	Sediment	AA79542
VC1	1600 ± 70	**1442** (1301-1568, 1588-1591)	Peat	LP-1729
VC2	3550 ± 70	**3769** (3580-3931, 3942-3968)	Peat	LP-1728
VC4	3633 ± 41	**3878** (3720-3985, 4052-4061)	Gyttja	AA79541
VC3	3860 ± 43	**4197** (4007-4032, 4081-4358, 4366-4407)	Gyttja	AA79540
AB21	2280 ± 70	**2225** (2011-2016, 2041-2353)	Gyttja	LP-1741
AB31	2525 ± 70	**2548** (2356-2724)	Gyttja	LP-1745
AB46	2820 ± 70	**2874** (2750-3041, 3047-3062)	Gyttja	LP-1740
AB50	2720 ± 70	**2786** (2504-2530, 2537-2594, 2614-2637, 2696-2955)	Gyttja	LP-1738
AB56	3110 ± 70	**3258** (3007-3013, 3033-3050, 3061-3409, 3426-3441)	Gyttja	LP-1737
AB67	3490 ± 70	**3696** (3482-3539, 3551-3873)	Gyttja	LP-1752
AB77	3812 ± 40	**4119** (3976-4248, 4275-4281)	Gyttja	AA-81415
Stingl and Garleff [18]	4080 ± 75	**4514** (4267-4270, 4288-4728, 4751-4820)	Gyttja	

[a] VC = Vega de la Cueva; AB = Agua Buena.
[b] AA = AMS radiocarbon dates; LP = conventional radiocarbon dates.
[c] CALIB 5.0.2 [28]. Median probability age and two sigma range given in parentheses.
[d] AA = University of Arizona AMS Facility; LP = LATYR, CONICET/Universidad Nacional de La Plata.

Some of the organic-rich levels were sampled for radiocarbon dating. All ^{14}C dates obtained were calibrated with CALIB 5.0.2 [23, 24] using the Southern Hemisphere calibration dataset [25]. The AB sequence was sampled every 10 cm in average with higher resolution in selected sections. Organic and carbonate contents of the sediment were determined by loss-on-ignition at burning temperatures of 550°C during four hours, and 900 °C during two hours, respectively [26]. Standard laboratory techniques were used for pollen extraction (KOH, HCl, HF, and acetolysis; [27]). Only samples from the organic sedimentary levels were analyzed since other samples usually contain low pollen concentration with bad preservation. Pollen grains were identified at 1000x magnification to the lowest taxonomic level possible based on modern pollen collections at the Laboratorio de Paleoecología y Palinología (UNMdP) and published atlases [28, 29]. Counts ranged from 200 to 350 terrestrial pollen grains (excluding aquatics and wetland taxa). Percentages of terrestrial taxa were based on the total sum of pollen grains from shrubs and herbs excluding wetland (Cyperaceae, Typha, Ranunculaceae) and aquatic taxa (Myriophyllum, Potamogeton/Triglochin). Wetland and aquatic taxa percentages were calculated upon a sum of all pollen taxa. Pollen calculations and diagrams were made on TGView-Tilia 1.0.1 program [30, 31] and only terrestrial, wetland and aquatic taxa corresponding to local communities are shown.

3. Results

3.1. Sedimentary record

Figure 2 displays the main geomorphological units of the area. Two stratigraphic sequences were analyzed, Agua Buena (AB) and Vega de la Cueva (VC). AB is located in the alluvial fan of the Bayo creek, one of the tributaries of the Atuel River. This mid-late Holocene unit connects laterally with a wetland containing a sedimentary infilling whose lower part is recorded in the VC sequence and became active around 3800 cal yr BP. Eolian deposits most probably of recent age cover the surface of the alluvial fan. The geometry and location of the alluvial fan facies was conditioned by a diamicton deposit (Figure 2) that may have forced the development of the distal facies located upstream with respect to the Atuel River flow (west-east).

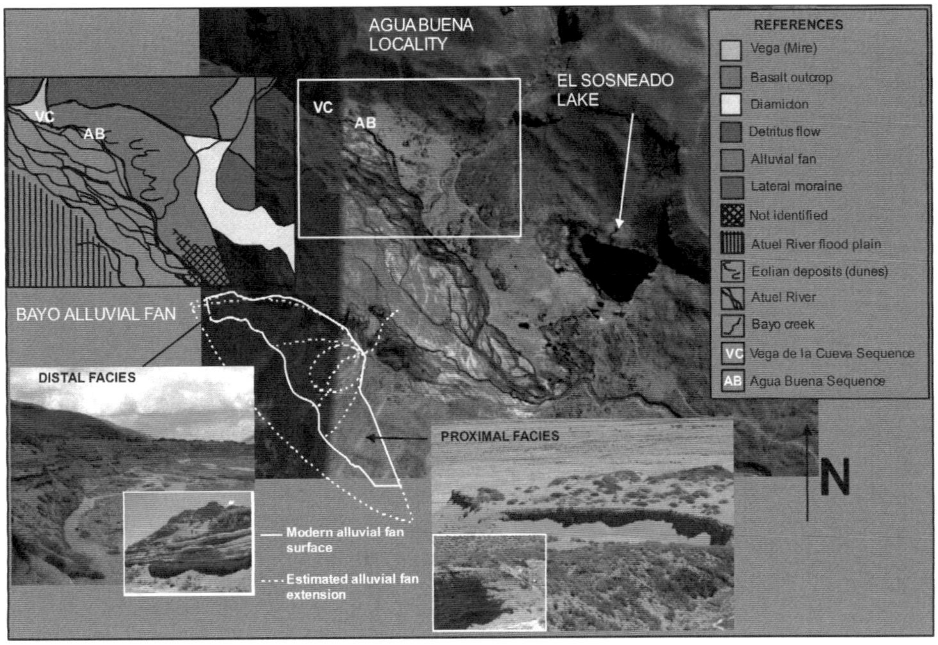

Figure 2. Geomorphological units identified in the Agua Buena locality.

3.1.1 Agua Buena sequence (AB). This 8 m sedimentary sequence has been exposed by erosion of the Atuel River along the distal part of the Arroyo Bayo alluvial fan (Figure 3a). The chronology of the sequence was established with seven radiocarbon samples at the uppermost 2 cm of organic-rich levels, yielding ages from 4100 and 2200 cal yr BP at 7.6 m and 2 m sediment depth, respectively

(Table 1). The basal section is composed of tabular cross-stratified gravels that pass abruptly into alternating layers of fine sand, clay and silt along with diatomite-like deposits and organic –rich levels (10 to 60 % organic matter (OM)). Since the pollen content of sands, clays and silts was very low, only the pollen records of the organic levels are shown. Poaceae (grasses) presented high values (15-94%) compared with the modern pollen of grass steppes registered in the area (<15%); hence it was considered as a component representing mainly the local vegetation. The record is dominated by wetland-aquatic taxa like Cyperaceae, Juncaceae and Myriophylum with values between 30 and 98%. The halophytes (Chenopodiaceae) reach 90% in some cases. Combined analysis of the stratigraphic and pollen records allowed distinguishing two distinctive sections. The lower section, between ca. 4100 and 2800 cal yr BP (8 to 5 m depth, respectively), shows thicker organic levels (9-13 cm) with higher OM content (15-60%) and elevated wetland taxa percentages. The upper section, between ca. 2800 and ca. 2300 cal yr BP (upper 5 m), shows thinner and less developed organic levels (3-7 cm) with lower OM content (10-15%), the presence of diatomite-like levels and dominance of interstratified horizontal layers of medium-coarse sands, fine sands, and silty clays. The pollen record of the upper section exhibits a decrease in wetland taxa and a rise in Poaceae and halophytes.

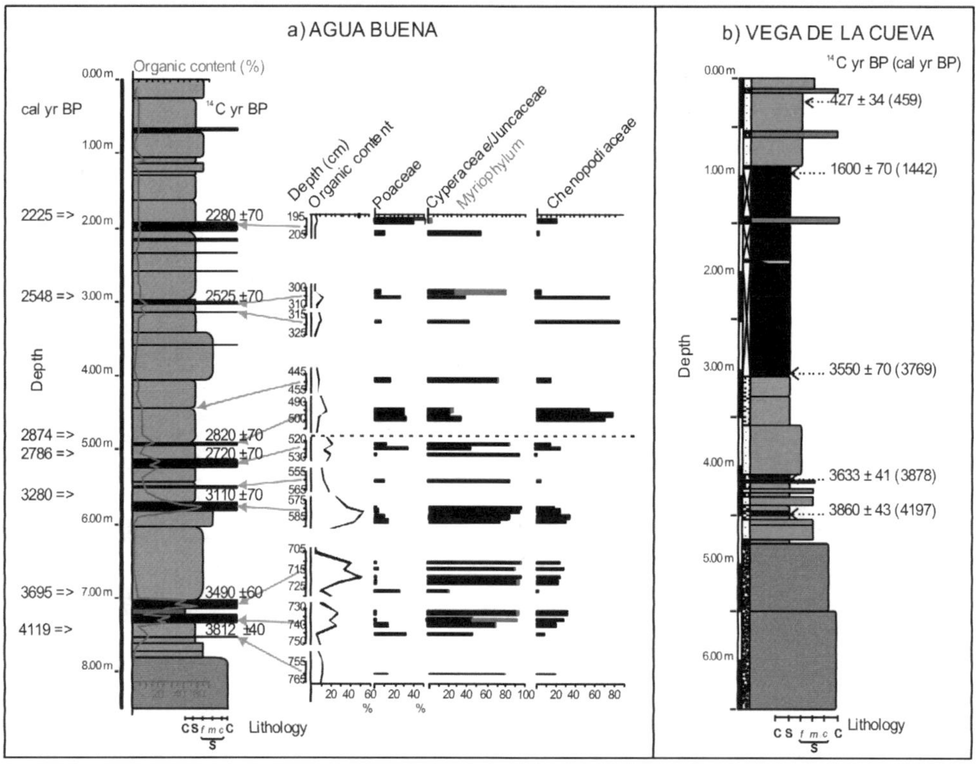

Figure 3. a) Stratigraphic and pollen profiles of the Agua Buena (AB) site. The blue curve indicates organic matter percentages. **b)** Vega de la Cueva (VC) stratigraphic section. Organic layers (peat-like deposits) are pictured in black. Lithologies are as follows: C, clay; S, silt; Sf, fine sand; Sm, medium sand; Sc, coarse sand; G, gravel.

3.1.2 Vega de la Cueva sequence (VC). The 6.5 m thick studied section is located in a wetland developed by a confined creek in the immediate vicinity of the northwestern tip of the Bayo creek alluvial fan (Figures 2 and 3). The chronology of the sequence was developed with five radiocarbon ages in samples from the top of the organic levels yielding ages of 4200 and 500 cal yr BP at 4.5 m and 0.22 m sediment depth, respectively (Table 1). The 2 m thick basal part is composed of tabular cross-stratified gravels with imbricate pebbles. Upwards it continues into levels of rippled and horizontal laminated sands, clays and silts including two thin layers (5 cm) with high OM content dated at 4100 and 3900 cal yr BP, respectively. An almost continuous deposit of organic matter with coarse plant debris and low clastic content developed between 3.1 and 0.9 m including a 1 cm thick, volcanic ash layer at 1.9 m and a fluvial lens. The minimum age for the beginning of this organic deposit is 3800 cal yr BP, continuing almost uninterrupted until 1400 cal yr BP. The top of this level shows a thin veneer of white silty diatomaceous sediments. The uppermost 1.8 m of the VC sequence is composed of layered deposits of fine to medium sands topped by a 5-10 cm tabular gravel bed.

4. Discussion

The basal tabular cross-stratified gravel beds observed at the VC profile are similar to modern deposits of the Atuel River suggesting a braided fluvial depositional environment. The location of the riverbed, and presumably the hydrological conditions of the basin, changed sometime before 4100 cal yr BP. The AB record corresponds to the distal facies of the Arroyo Bayo alluvial fan, starting the aggradation process prior to ca. 4100 cal yr BP. The organic-rich levels can be considered as limnic events in the sense of the Soil Taxonomy [32] or gyttja levels [33]. They are the result of the accumulation of organic matter in highly productive environments with almost permanent water saturation. Cyperaceae, Juncaceae and Poaceae dominate the communities that developed in these "vegas" whereas the high values of Myriophyllum indicate the development of small water bodies or streams. These deposits occurred more regularly and lasted longer between 4100 and 2800 cal yr BP indicating more stable conditions coinciding with a decrease in sedimentation rates during aggradation in the alluvial fan environment. After 2800 cal yr BP, the organic content is comparatively lower with increasing sedimentation rates that are indicative of higher fluvial discharges. In addition, the increase of Chenopodiaceae and Poaceae, the relatively lower values of wetland taxa and the lower productivity in AB also suggest more dynamic environments with limited water availability for the development of vegas. The middle section of the VC profile is similar to the organic levels at AB and presumably indicates the continuous development of wetlands at least between 3800 and 1400 cal yr BP. Sand lenses and small gravel levels suggest the occurrence of small water streams during its formation. Finally the upper section with sand banks is correlated with the distal facies of the Arroyo Bayo alluvial fan.

The results of this study are not conclusive concerning the occurrence of a neoglacial advance during the mid-Holocene. Furthermore, the two studied sections suggest that the deposits do not correspond neither to an alluvial terrace of the Atuel River [14, 15, 16] nor to lake deposits of a big lake formed by the damming of the river [17]. Instead, the lithostratigraphic arrangement and geometry of the deposits along with the geomorphological evidences point to distal facies of an alluvial fan with some intervals of wetland formation and simultaneous development of "vega" environments in its flanks. This information is essential when interpreting the pollen records of these sequences as well as to evaluate both their representativeness and potential for the reconstruction of local and/or regional vegetation. Further studies will include this information in order to reconstruct the environmental history during the mid-late Holocene in the area and discuss the paleoclimatic significance of these records.

Acknowledgements

This study was financed by CONICET grant PIP-5819 to M. Zarate and UNMdP grant EXA-431/08 to M. Paez. D. Navarro and A. Mehl were supported by Postgraduate Fellowships from CONICET. We would like to thank Adolfo Gil, Gustavo Neme, Leandro Rojo and Alejandra Guerci at Departamento de Antropología, Museo de Historia Natural de San Rafael for their fieldwork assistance and fruitful discussions. We would also like to thank Daniel Ariztegui and Cathy Whitlock for the helpful and detailed reviews, and Caroline Cléroux and Thorsten Kiefer for the edition work. DN is very grateful to PAGES and the US National Science Foundation for financial support to assist to PAGES 1st YSM and 3rd OSM, to the PAGES' staff for organizing the meetings and to Cathy Whitlock for continuous encouragement and scientific support.

References

[1] Lamy F, Hebbeln D and Wefer G 1999 High resolution marine record of climatic change in mid-latitude Chile during the last 28,000 years based on terrigenous sediment parameters *Quat. Res.* **51** 83-93
[2] Markgraf V (ed) 2001 *Interhemispheric Climate Linkages* (San Diego, CA: Academic Press) 454 pp
[3] Jenny B, Valero-Garcés B L, Villa-Martinez R, Urrutia R, Geyh M and Veit H 2002 Early to mid-Holocene aridity in central Chile and the southern westerlies: The Laguna Aculeo record (34°S) *Quat. Res.* **58(2)**160-70
[4] Lamy F, Kaiser J, Ninnemann U, Hebbeln D, Arz H W and Stoner J 2004 Antarctic timing of surface water changes of Chile and Patagonian ice sheet response *Science* **304** 1959-62
[5] Mancini M V, Paez M M, Prieto A R, Stutz S, Tonello M and Vilanova I 2005 Mid-Holocene climatic variability reconstruction from pollen records (32°-52°S, Argentina) *Quat. Int.* **132** 47-59
[6] Whitlock C, Moreno P I and Bartlein P 2007 Climatic controls of Holocene fire patterns in southern South America *Quat. Res.* **68** 28-36
[7] Mancini M V, Prieto A R, Paez M M, Schäbitz F 2008 Late Quaternary vegetation and climate of Patagonia *The Late Cenozoic of Patagonia and Tierra del Fuego (Developments in Quaternary Sciences vol 11)* ed J Rabassa (Amsterdam: Elsevier) pp 351-367
[8] Paez M M, Navarro D, Rojo L and Guerci A In press Vegetación y paleoambientes durante el Holoceno en Mendoza *Paleoambientes y Ocupaciones Humanas del Centro-Oeste de Argentina Durante el Pleistoceno Tardío – Holoceno* eds M A Zarate *et al* (Buenos Aires: Editorial Sociedad Argentina de Antropología)
[9] Navarro D, Whitlock C, Paez M M and Zarate M A Examining the mid-late Holocene environments of west-central Argentina *3rd PAGES Open Science Meeting: Retrospective Views on Our Planet's Future* 8-11 July 2009 Corvallis USA p 104
[10] Garreaud R D, Vuille M, Compagnucci R and Marengo J. 2009. Present-day South American climate. *Palaeogeogr Palaeoclimatol Palaeoecol* **281(3-4)** 180-95
[11] Hoffman J A J 1975 *Atlas climatológico de América del Sur* (Geneve: OMM-WMO-UNESCO)
[12] Prohaska F 1976 The climate of Argentina, Paraguay and Uruguay *Climates of Central and South America (World Survey of Climatology vol 12*) ed W Schwerdtfeger (Amsterdam: Elsevier) pp 13-112
[13] Cabrera A L 1971 Fitogeografía de la República Argentina *Boletín de la Sociedad Argentina de Botánica* **14(1-2)** 1-50
[14] Stingl H and Garleff K 1978 Gletscherschwankungen in den subtropisch-semiariden Hochanden Argentiniens *Z. Geomorph. N.F. Suppl. Bd.* **30** 115-31
[15] Stingl H and Garleff K 1985 Glacier variations and climate of the late Quaternary in the subtropical and Mid-Latitude Andes of Argentina *Z. Gletzcherkd. Glazialgeol.* **21** 225-28
[16] Garleff K and Stingl H 1994 Reply to Gosse and Evenson: Reinterpretation of the evidence for a significant neoglacial ice advance in the Rio Atuel valley, Mendoza Province, Argentina *Z. Geomorph. N.F.* **38(3)** 339-42
[17] Gosse J C and Evenson E B 1994 Reinterpretation of the evidence for a significant mid-Holocene ice advance in the Rio Atuel valley, Mendoza Province, Argentina *Z. Geomorph. N.F.* **38(3)** 327-38
[18] Cobos D R and Boninsegna J A 1983 Fluctuations of some glaciers in the upper Atuel River basin, Mendoza, Argentina *Quaternary of South America and Antarctic Peninsula* **1** 61-82
[19] Le Quesne C, Acuña C, Boninsegna J A, Rivera A and Barichivich J 2009 Long-term glacier variations in the Central Andes of Argentina and Chile, inferred from historical records and

tree-ring reconstructed precipitation *Palaeogeogr Palaeoclimatol Palaeoecol* **281(3-4)** 334-344

[20] Miall A D 2006 *The Geology of Fluvial Deposits: Sedimentary Facies, Basin Analysis and Petroleum Geology* (Berlin: Springer-Verlag) 582 pp

[21] Reading H G 1996 *Sedimentary Environments: Processes, Facies and Stratigraphy* Third Edition (Malden: Blackwell Science) 688 pp

[22] Catt J A 1990 *Paleopedology Manual* Quat. Int. 6 (Oxford: Pergamon press) 95 pp

[23] Stuiver M and Reimer P J 1993 Extended 14C database and revised CALIB radiocarbon calibration program *Radiocarbon* **35** 215-30

[24] Stuiver M, Reimer P J and Reimer R W 2005 *CALIB 5.0* [http://calib.qub.ac.uk/calib/].

[25] McCormac F G, Hogg A G, Blackwell P G, Buck C E, Higham T F G and Reimer P J 2004 SHCal04 Southern Hemisphere Calibration 0-11.0 cal Kyr BP *Radiocarbon* **46** 1087-92

[26] Heiri O, Lotter A F and Lemcke G 2001 Loss on ignition as a method for estimating organic and carbonate content in sediments: reproducibility and comparability of results *J. Paleolimnol.* **25** 101–10.

[27] Faegri K and Iversen J 1989 *Textbook of Pollen Analysis* 4th ed by Faegri K, Kalland P E and Krzywinski K (Chichester: J. Wiley & Sons) 314 pp

[28] Heusser C 1971 *Pollen and Spores of Chile. Modern Types of the Pteridophytas, Gimnospermae and Angiospermae* (Tucson: The University of Arizona Press)

[29] Markgraf V and D'Antoni H L 1978 *Pollen flora of Argentina* (Tucson: The University of Arizona Press) 201 pp

[30] Grimm E C 1987 CONISS: a FORTRAN 77 program for stratigraphically constrained cluster analysis by the method of incremental sum of squares *Comput. Geosci.* **13** 13-35

[31] Grimm E C 2007 TGView-Tilia program Version 1.0.1. Illinois State Museum. Research and Collections Center. Springfield, Illinois USA.

[32] Soil Survey Staff 2003 *Keys to Soil Taxonomy* United States Department of Agriculture, Natural Resources Conservation Service. 9th Edition 334 pp

[33] Fox C A 1985 Micromorphological characterization of histosols *Soil micromorphology and soil classification (SSSA Special Publ. no. 15)* ed L A Douglas and M L Thompson (Madison: Soil Sci. Soc. Am.) 216 pp

IOP Conf. Series: Earth and Environmental Science **9** (2010) 012015 doi:10.1088/1755-1315/9/1/012015

^{230}Th/U-dating of a late Holocene low uranium speleothem from Cuba

C Fensterer[1], D Scholz[2,3], D Hoffmann[2,4], A Mangini[1] and J M Pajón[5]

[1] Forschungsstelle Radiometrie, Heidelberg Academy of Sciences, Im Neuenheimer Feld 229, 69120 Heidelberg, Germany

[2] School of Geographical Sciences, University of Bristol, University Road, BS8 1SS, Bristol, United Kingdom

[3] Institute for Geosciences, University of Mainz, Johann-Joachim-Becher-Weg 21, 55128 Mainz, Germany

[4] Centro Nacional de Investigatión sobre la Evolución Humana (CENIEH), Paseo Sierra de Atapuerca s/n, 09002 Burgos, Spain

[5] Department of Archaeology, Cuban Institute of Anthropology, Amargura No. 203, e/n Habana y Aguiar, Ciudad de La Habana, CP: 10 100, Cuba

E-mail: `Claudia.Fensterer@iup.uni-heidelberg.de`

Abstract. We present 22 U-series ages for a stalagmite from north-western Cuba based on multi-collector inductively coupled plasma mass spectrometry (MC-ICPMS) and thermal ionisation mass spectrometry (TIMS). Our results reveal that the stalagmite continuously grew within the last ~1400 a. Low uranium content of the sample and thus, extremely low ^{230}Th concentrations limit the precision and accuracy of ^{230}Th/U-dating by TIMS. Samples measured by MC-ICPMS show a high variability of ^{232}Th content along the growth axis with some sections significantly affected by initial ^{230}Th from a detrital phase. An a-priori bulk earth ratio for (^{238}U/^{232}Th) cannot be used to accurately account for this initial ^{230}Th. Using an age model based on the ^{230}Th/U ages determined on samples with low or negligible ^{232}Th concentration, we find that the (^{238}U/^{232}Th) activity ratio of the detrital phase is an order of magnitude larger than the bulk earth value, indicating the importance of an accurately determined correction factor.

1. Introduction

The Caribbean represents an important area to obtain information about past natural climate variability in the tropical region and its relationship to climate forcing mechanisms, such as changes in insolation. Speleothems are important archives for high resolution climate reconstructions and can be accurately dated by U-series disequilibrium dating methods [1]. The two most precise analytical techniques for the determination of U and Th isotope ratios are multi-collector inductively coupled plasma mass spectrometry (MC-ICPMS) and thermal ionisation mass spectrometry (TIMS) with precisions between per mill and percent, depending on the U content and the age of the sample [2]. Applicability of ^{230}Th/U-dating on speleothems is based on the prerequisite that no initial ^{230}Th (i.e., not produced by in situ decay of ^{234}U and ^{238}U) is incorporated when the $CaCO_3$ forms and that the crystal remains a closed system after deposition [3]. However, in practice small quantities of initial ^{230}Th are often incorporated during precipitation of speleothem $CaCO_3$. Usually, the ^{232}Th concentration is used as a proxy for initial ^{230}Th [4], and the ^{230}Th/^{232}Th activity ratio (henceforth activity ratios are indicated by parentheses, e.g., (^{230}Th/^{232}Th)) indicates whether a correction needs to be applied. As a threshold for mass spectrometric analyses an activity ratio between 100 and 300 has been suggested [5, 6]. There are different methods to estimate the extent of contamination with initial ^{230}Th: (1) direct measurement of the isotopic composition of the contaminating phase using an isochron technique on multiple sub samples [4]. (2) Assuming an a-priori estimate for the isotopic composition of the contaminating phase [3, 7, 8], e.g., the bulk earth ^{232}Th/^{238}U concentration ratio of the upper crust of 3.8 [9] and secular equilibrium for ^{230}Th, ^{234}U and ^{238}U. Depending on the (^{230}Th/^{232}Th) of the sample the correction may have a significant effect on the ^{230}Th-^{234}U-^{238}U activity ratios, and it is obvious that the accuracy of the corrected ratios depends on the accuracy of the correction factor. Using the a-priori estimate, the effect of propagating the assumed uncertainty of the isotopic composition of the contaminating phase through the age calculation may also be quite severe [8]. Thus, in case that a correction for detrital contamination is applied, precise direct measurement of the isotopic composition of the contaminating phase is required. The compostition of detrital phases can be very different from the a-priori assumed bulk earth value, e.g., due to partial leaching effects, Th adsorbtion, alpha recoil or non silicate origin of the detrital phase prior to deposition [8].

Here we present U-series data of a stalagmite from western Cuba that has an extremely low ^{230}Th concentration. Thus, the effect of potential contamination with initial ^{230}Th is very large. Since no distinct growth layers are visible, isochrons are not applicable and an alternative method is used to constrain the (^{238}U/^{232}Th) of the contaminating phase. First, a robust age model using uncorrected (^{230}Th/^{238}U) and (^{234}U/^{238}U) ratios is derived from sub samples with low ^{232}Th concentration measured with high precision MC-ICPMS. Then the dating results of samples with elevated ^{232}Th concentration are compared with this age model to estimate the (^{238}U/^{232}Th) of the contaminating phase. This enables the derivation of a sample specific (^{230}Th/^{232}Th) for the correction.

2. Location, sample and methods

Dos Anas cave (Pinar del Rio, north-western Cuba, 22° 23' N, 83° 58' W) has a total length of ca. 14 km and is part of the Majaguas Cantera cave system [10]. Stalagmite Cuba Grande (CG) grew in Dos Anas cave at an elevation of ~100 m above sea level and ~150 m under the surface. The sampling location, which is located in the Salon de la Cimitarra, lies in a distance of ca. 1.5 km from the cave entrance. The 720 mm long stalagmite sample consists of pure white calcite, as confirmed by X-ray diffraction analysis. No distinct growth layers are visible (Fig. 1).

22 MC-ICPMS and TIMS U-series ages were determined along the growth axis of CG. Ten MC-ICPMS measurements were performed using a ThermoFinnigan Neptune MC-ICPMS at the Bristol Isotope Group, and twelve analyses were performed on a Finnigan MAT 262 RPQ TIMS at the Heidelberg Academy of Sciences. The key difference between both methods is the transfer efficiency of isotopes, which is $\ll$1% with TIMS, but ~1% with MC-ICP-MS [2]. For young samples with low uranium content, the MC-ICPMS technique thus enables both a higher precision of ^{230}Th/U-dating and to use smaller sample sizes. For both methods sample masses between 0.5 and 1 g were used for one analysis. The average span of time included in one sample should, thus, be the same for MC-ICPMS and TIMS. Due to the smaller sample size required for analysis, however, the precision of the MC-ICPMS ages is much higher. Chemical preparation of the samples follows the methods described in Hoffmann et al. (2007) [11] and Hoffmann (2008) [12] for MC-ICPMS and in Scholz et al. (2004) [13] for TIMS. Different U and Th spikes were used in both laboratories. The calibration of both spikes and an inter-laboratory comparison is described in Hoffmann et al. (2007) [11]. The comparison did not show any significant differences for the two spike solutions. Ages were initially calculated based on uncorrected (^{230}Th/^{238}U) and

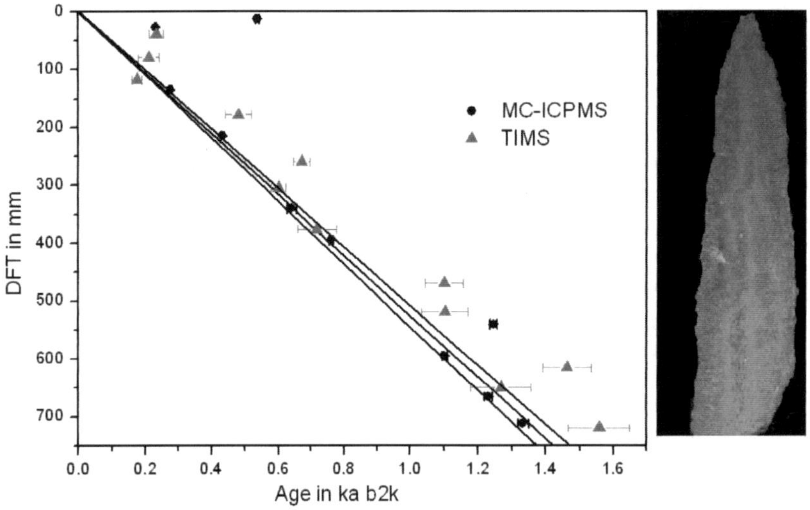

Figure 1. (left) Ages determined by the MC-ICPMS (black circles) and the TIMS method (red triangles), without any detrital correction. Ages are given in ka b2k as a function of distance from top (DFT) in mm. The linear age model for the MC-ICPMS data yields a mean growth rate of 526 μm/a (black line). The corresponding 95 % confidence limits are also shown. The two uppermost and the sample at 540 mm DFT have not been considered for the linear fit. (right) Picture of stalagmite CG.

($^{234}U/^{238}U$) and corrected ratios, using a bulk earth ($^{238}U/^{232}Th$) of 0.8 where decay products of detrital ^{238}U are assumed to be in secular equilibrium. The quoted age uncertainties include uncertainties from the detrital correction. Details about uncertainties of MC-ICPMS can be found in Hoffmann et al. (2007) [11]. All uncertainties are quoted at 95 % confidence level.

3. Results

All $^{230}Th/U$ dating results are shown in Fig. 1, and a table with all data can be found in the supplementary information. Ages are given in ka before the year 2000 (ka b2k). The U content of the samples ranges from 90 to 200 ng/g. ^{230}Th concentrations vary between 0.01 and 0.06 pg/g of sample. Analysis of such small concentrations of ^{230}Th is more precise using the MC-ICPMS technique, hence the focus of further examinations is on the MC-ICPMS data. Seven ages determined by MC-ICPMS suggest nearly linear growth during the last ~1370 years. The calculated age model using a linear fit has a slope of $0.00190(\pm 2\cdot 10^{-5})$ ka/mm, which corresponds to a growth rate of $526(\pm 11)\,\mu$m/a. The fit was forced through the origin since the dripsite was active when the stalagmite was removed from the cave. Three ages at 13, 27 and 540 mm distance from top (DFT) are not in stratigraphic order and have, thus, not been considered for this fit. The ages determined by the TIMS method generally exhibit older ages than the MC-ICPMS data and a larger scatter (Fig. 1).

Fig. 2 shows the ages determined by MC-ICPMS in comparison to the ($^{230}Th/^{232}Th$) of the samples. The three samples at 13, 27 and 540 mm DFT show a substantially lower ($^{230}Th/^{232}Th$) than the other samples suggesting a higher degree of detrital contamination. Since these three ages do not plot on the straight line determined by the less contaminated ages, this suggests that the applied correction using a ($^{238}U/^{232}Th$) of 0.8 is inadequate to account for the effect of detrital contamination. The samples at 13, 27 and 540 mm DFT show a ($^{230}Th/^{232}Th$) ratio

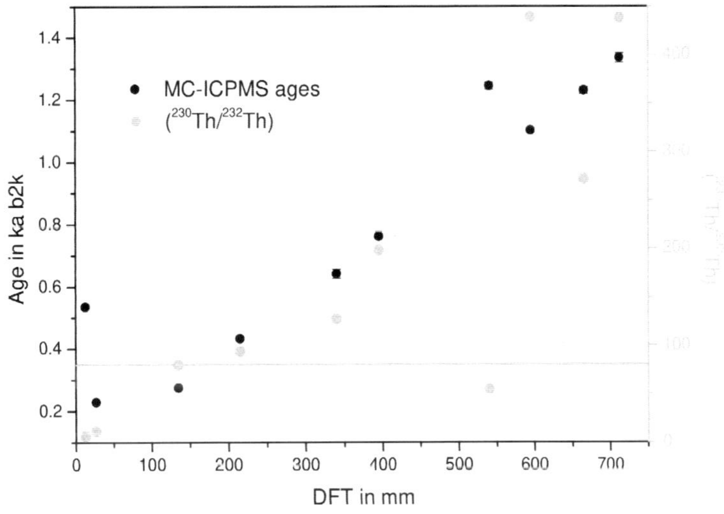

Figure 2. MC-ICPMS dating results (black circles) and (^{230}Th/^{232}Th) (green circles). The three ages at 13, 27 and 540 mm DFT are not in stratigraphic order, and the corresponding samples have a significantly lower (^{230}Th/^{232}Th) than 80 (green line).

below 80, which we use as a threshold to identify results that require a detritus correction.

4. Discussion

A strict constraint for the age of sub-samples taken along the growth axis of a stalagmite is, that ages must become progressively older from top to bottom [2, 8, 14, 15]. Three ages at 13, 27 and 540 mm DFT are not in stratigraphic order and therefore do not fulfill this criterion (Figs. 1 and 2). Assuming that the true age-depth relationship for stalagmite CG is described by the linear age model determined using the samples with low ^{232}Th content (Fig. 1), the (^{238}U/^{232}Th) for the correction of the samples with elevated ^{232}Th can be estimated. We find that the activity ratios of the samples at 13, 27 and 540 mm DFT need to be corrected with (^{238}U/^{232}Th) ratios of 6.1, 8.7 and 9.6, respectively, to obtain corrected ages in agreement with the linear age-depth model. The mean of the three (^{238}U/^{232}Th) ratios is 8($\pm$4) (2σ standard deviation of the average value), almost an order of magnitude bigger than the bulk earth factor, which is normally used to account for detrital contamination. This (^{238}U/^{232}Th) is now used to recalculate all ages determined by MC-ICPMS and TIMS. The results are shown in Fig. 3.

The corrected age of the uppermost sample dated by the MC-ICPMS method is out of range using the mean correction factor of 8. This is due to the fact that in this case the assumed ^{230}Th concentration of the detrital phase is larger than the total ^{230}Th concentration. Thus, the corrected concentration of ^{230}Th becomes negative, and the ^{230}Th/U age equation cannot be solved. Instead we use the maximum value to maintain positive results and the minimum value of the confidence range of the correction factor (6 and 4 respectively) to estimate the age range for this sample.

The corrected TIMS ages are still not in stratigraphic order and in general exhibit older ages than the results from MC-ICPMS. This suggests that the scatter in the TIMS data is not solely

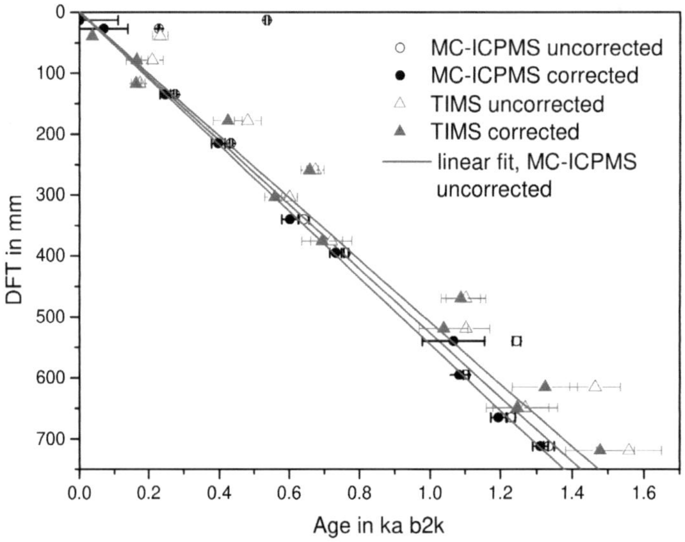

Figure 3. MC-ICPMS dating results, corrected with a ($^{238}U/^{232}Th$) of 8 (filled black circles) and without correction (open black circles). TIMS ages are shown as filled red triangles for the corrected resulst, black open triangles for uncorrected. The age model determined by the uncorrected MC-ICPMS data is shown as a gray line.

due to contamination with initial ^{230}Th. Differences in spike calibration can be excluded as the reason for the apparently too old TIMS ages since the inter-laboratory comparison did not show significant differences [11]. In contrast, we speculate that a relatively large contribution of the background intensity to the ^{230}Th signal is the reason for this scatter, which also limits the precision that can be achieved by TIMS. Taking into account the relatively large uncertainties of the TIMS ages, the TIMS age model would be in agreement with the age model based on MC-ICPMS results.

5. Conclusions

Results from 22 MC-ICPMS and TIMS $^{230}Th/U$ age determinations on a Holocene low U stalagmite from north-western Cuba show that MC-ICPMS measurements generally yield smaller errors, which is due to the higher sensitivity especially for ^{230}Th and hence better counting statistics. Furthermore, a background contribution for TIMS measurements becomes significant for samples with low U content due to extremely small count rates.

Initial ^{230}Th has a significant effect on such samples and requires an accurately determined detrital correction factor. Since isochrons are not applicable, an alternative approach was used: subsamples with negligible detritral contribution allow derivation of an age model, which constrains the detrital ($^{230}Th/^{232}Th$) of the other samples. We find a ($^{230}Th/^{232}Th$) that is an order of magnitude larger than the bulk earth value commonly applied for detrital correction. A similarly elevated initial ($^{230}Th/^{232}Th$) has also been reported by Beck et al. (2001) [14] for a stalagmite from the Bahamas and may result from incorporation of carbonate dust during formation.

The sample specific (^{238}U/^{232}Th) can be used to correct results of samples with significantly higher ^{232}Th concentration. For the Cuban stalagmite CG we find a value of 8(±4) to correct for detrital components.

Acknowledgements
We kindly thank Thomas Felis, Cyril Giry and Michael Zuther from the University of Bremen for the X-ray diffraction measurements. We like to thank Gabriel García, Paolo Terzan and Ana Abraham for assistance in the field in Cuba. We also thank the anonymous reviewer for valuable suggestions which helped to improve the manuscript. The MC-ICPMS measurements were accomplished by Claudia Fensterer during a research stay at the Bristol Isotope Group at the University of Bristol which was funded by the DFG SPP 1266 Interdynamik and kindly approved by Tim Elliot. Funding to attend the PAGES 1st YSM in 2009 was also provided by Interdynamik. Denis Scholz was funded by the DFG (SCHO 1274/1-1).

References

[1] Henderson G M 2006 Caving in to new chronologies *Science* **313** 620-622

[2] Scholz D and Hoffmann D 2008 ^{230}Th/U-dating of fossil corals and speleothems *Quaternary Science Journal* **57/1-2** 52-76

[3] Ivanovich M and Harmon R S 1993 *Uranium Series Disequilibrium: Applications to Environmental Problems* (Oxford: Clarendon Press)

[4] Ludwig K R and Titterington D M 1994 Calculation of ^{230}Th/U isochrons, ages, and errors *Geochimica et Cosmochimica Acta* **58** 5031-5042

[5] Li W X, Lundberg J, Dickin A P, Ford D C, Schwarcz H P, McNutt R and Williams D 1989 High precision mass-spectrometric uranium-series dating of cave deposits and implicatinos for palaeoclimate studies *Nature* **339** 534-536

[6] Richards D A and Dorale J A 2003 Uranium-series chronology and environmental applications of speleothems *Reviews in Mineralogy & Geochemistry* **52** 407-460

[7] Dorale J A, Edwards R L, Ito E and Gonzalez L A 1998 Climate and vegetation history of the midcontinent from 75 to 25 ka: a speleothem record from Crevice Cave, Missouri, USA *Science* **282** 1871-1874

[8] Hellstrom J C 2006 U-Th dating of speleothems with high initial ^{230}Th using stratigraphical constraint *Quaternary Geochronology* **1** 289-295

[9] Wedepohl H K 1995 The composition of the continental crust *Geochimica et Cosmochimica Acta* **59** 1217-1232

[10] Pajón J M., Hernández I, Ortega F and Macle J 2001 Periods of wet climate in Cuba: evaluation of expression in karst of Sierra de San Carlos in *Interhemispheric Climate Linkages* ed V Markgraf (San Diego: Academic Press) ch 13 pp 217-226

[11] Hoffmann D L, Prytulak J, Richards D A, Elliott T, Coath C D, Smart P L and Scholz D 2007 Procedures for accurate U and Th isotope measurements by high precision MC-ICPMS *International Journal of Mass Spectrometry* **264** 97-109

[12] Hoffmann D L 2008 ^{230}Th isotope measurements of femtogram quantities for U-series dating using multi ion counting (MIC) MC-ICPMS *International Journal of Mass Spectrometry* **275** 75-79

[13] Scholz D, Mangini A and Felis T 2004 U-series dating of diagenetically altered fossil reef corals *Earth and Planetary Science Letters* **218** 163-178

[14] Beck J W, Richards D A, Edwards R L, Silverman B W, Smart P L, Donahue D J, Hererra-Osterheld S, Burr G S, Calsoyas L, Jull A J T and Biddulph D 2001 Extremely large variations of atmospheric C-14 concentration during the last glacial period *Science* **292** 2453-2458

[15] Musgrove M, Banner J L, Mack L E, Combs D M, James E W, Cheng H and Edwards R L 2001 Geochronology of late Pleistocene to Holocene speleothems from central Texas: implications for regional paleoclimate *Geological Society of America Bulletin* **113** 1532-1543

IOP Conf. Series: Earth and Environmental Science **9** (2010) 012016 doi:10.1088/1755-1315/9/1/012016

Quercus macrocarpa annual, early- and latewood widths as hydroclimatic proxies, southeastern Saskatchewan, Canada

J R Vanstone[1,2] and D J Sauchyn[1,2]

[1]Department of Geography, University of Regina, 3737 Wascana Pkwy, Regina, SK, S4S 0A2, Canada

[2] Prairie Adaptation Research Collaborative, University of Regina, #120 – 2 Research Drive, Regina, SK, S4S 7H9, Canada

jrvanstone@gmail.com

Abstract. Fluctuations in size of annual ring-widths of *Quercus* species suggest that environmental factors influence the size and density of vessels within the ring, either by acting as a limiting factor for growth or through fine tuning of the wood structure to environmental factors. The purpose of this study is to assess the potential of *Q. macrocarpa* to provide multiple dendroclimatic proxies for the Canadian Prairies, by investigating growth responses of annual, early- and latewood widths to regional climate variability. Results indicate that ring width chronologies, from southeastern Saskatchewan capture regional signals related to moisture and drought conditions. Correlations suggest that late-wood widths are more representative of annual ring-widths, than are early-wood widths, and are the best proxy of seasonal fluctuations in climate. Thus regression models that include latewood widths were able to account for more variance in the Palmer Drought Severity Index (PDSI) than when annual ring-widths are used as the only proxy. This study demonstrates that *Q. macrocarpa* can provide multiple dendroclimatic proxies for investigating large scale climatic fluctuations at annual and sub-annual time scales. It is novel in terms of sub-annual analysis of tree-rings in a region that previously lacked dendrochronological research.

1. Introduction

The Canadian Prairie region accounts for 82% of the cultivated land in Canada and is an important supplier of food for the global community, with agricultural commodities ranging from various grains, through legumes and oilseed, to both grain and grass-fed meat products [1, 2]. The predicted global warming of 1.4 to 5.8°C between 1900 and 2100 [3], demands understanding the vulnerability of the Prairie region to climate change and developing adaptation strategies [4]. Studying past climatic change puts current trends in perspective, reveals the mechanisms that govern the Earth's systems, and provides data to test predictive models and to evaluate the impact of climatic change on biological and geographical systems and processes [5]. Absolutely dated tree-ring chronologies are an important archive for evaluating both past and current natural climate variability occurring over centuries and millennia, as well as changes over recent decades and annual time frames [6]. As well, tree-rings offer the greatest potential for studying climatic and environmental variability at local and regional levels, because of the wide geographical distribution of suitable sites, high temporal resolution (annual or even seasonal), high replication factor and environmentally sensitive (*i.e.* accurate) nature of their growth patterns [7]. Most ring-width records for the Prairies are derived from trees growing along the margins of the Prairie Ecozone, [8] mostly because long-lived species are rare in the subhumid northern Great Plains [9]. Notable exceptions of long- tree-ring chronologies have been obtained from

Quercus species found scattered throughout the Prairies and Great Plains region [8, 10, 11, 12, 13]. Investigations of *Q. macrocarpa* (bur oak) in Nebraska and the Dakotas have indicated the longevity (most commonly 200 to 300 years old, with some samples reaching ages of over 700 years old) and moisture sensitivity of this species and its potential for drought investigation [8, 13].

Comparisons between patterns of annual layers of *Quercus* tree-rings and climate records suggest that these parameters may serve as potential climate proxies, whereby climatic factors influence the size and density of vessels within the ring, either by acting as a limiting factor for growth or through fine tuning of the wood structure to environmental (water) conditions [8, 13, 14]. Vessels are an advanced evolutionary trait of angiosperms that enable the trees to more efficiently transport water up the tree. Under normal growing conditions, ring-porous trees develop single or multiple rows of large conductive vessels in the spring (earlywood) and form smaller vessels (latewood) during the rest of the growing season [15]. In general, *Quercus* species have the ability to sustain stomatal conductance at low soil and leaf water potentials and have an inherently low capacity for water loss that contributes to their success in dry locations [16]. When water availability is limited during the growing season, disruptions in the physiological process that control cambial growth most often causes anomalous tissue development within the annual ring [17]. Diameter and density of the vessels have been related to vulnerability to water stress [18], with smaller, denser vessels, and narrower annual rings, predominantly being associated with an abnormal growth year (such as during floods or droughts) [17].

The size and density differences within each annual ring are imparted by differences in rates and durations of cell processes [19], resulting in the formation of early- and latewood. Latewood holds potential for studying summer drought; however, the application of latewood data to paleoclimatology has been limited by relatively few chronologies that have been developed and analyzed [19], and nearly all of these studies have examined the climate sensitivity of earlywood or latewood in coniferous species [19, 20, 21, 22], where ring boundaries are identified by examining the size and cell wall thickness of the tracheids. Therefore, with the highly distinguishable sub-annual differentiation of cellular structure/growth patterns found in ring-porous gymnosperms such as the *Quercus* species, the goal of this study is to assess the potential of *Q. macrocarpa* to provide multiple dendroclimatic proxies for the Canadian Prairies by investigating growth responses of annual, early- and latewood widths (RW, EW, and LW, respectively), to regional climate variability (*i.e.* climate variables such as temperature, precipitation and the Palmer Drought Severity Index, PDSI).

2. Methods

2.1 Tree-ring data

Cores and cross-sectional discs of bur oak trees were sampled from five sites within the Qu'Appelle Valley of southeastern Saskatchewan (Figure 1). As in conventional dendroclimatic studies, all samples were collected and prepared for analysis as described by Stokes and Smiley [23] and Fritts [6]. After surfacing, samples were scanned using an Epson® Expression 10000 flatbed scanner, at resolutions between 800 and 1200 dpi, depending on the clarity of the image required. Higher resolution images were required for enabling the distinction of narrow rings in periods of suppressed growth if the coarser resolution was deemed inefficient for this purpose. Dating, measuring and visual crossdating of annual rings were carried out using the program WinDendro™ Density (version 2008b). Early- and latewood width measurements were aided by the distinct size differentiation of early- and latewood vessels in bur oak samples. Early- and latewood measurements were based on a defined boundary of 40% of the minimum to maximum relative pixel density in the reflectance values, where the onset of latewood growth was noted as having more dense and compacted vessels being darker in nature. Some early- latewood boundaries were adjusted manually to correct for errors in the automatic detection process.

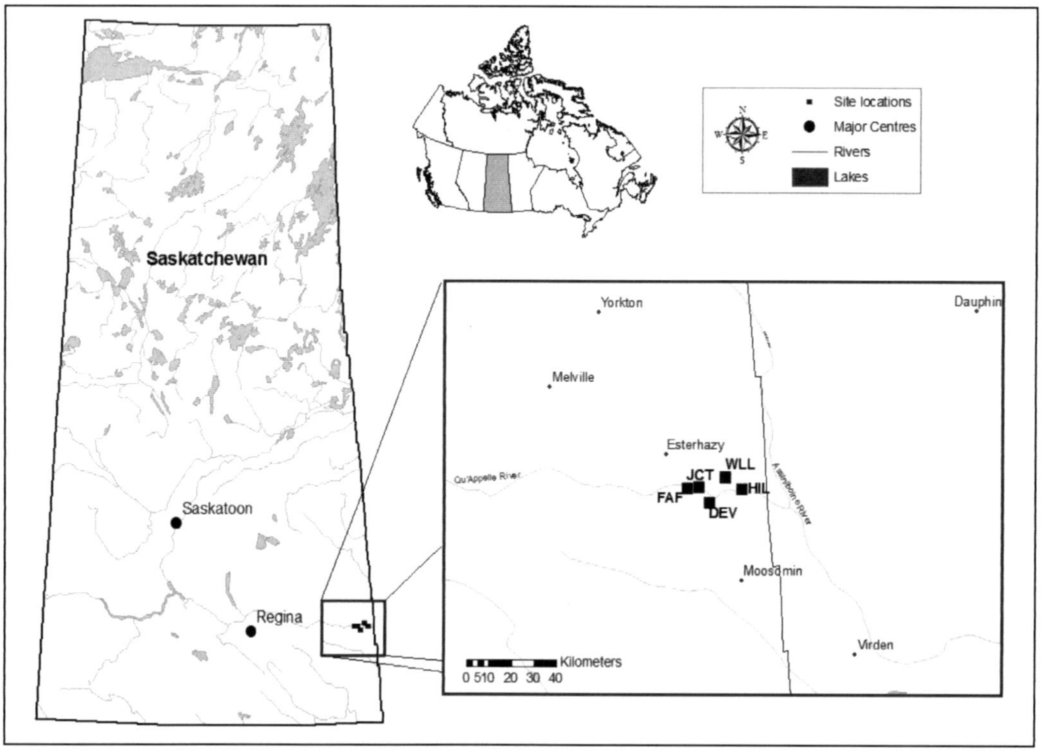

Figure 1. Chronology site locations in southeastern Saskatchewan.

COFECHA was used to verify crossdating and provide chronology statistics that describe the strength of intercorrelation between tree-ring samples at each site (Table 1) [24, 25]. Sites were individually crossdated and then combined to produce three regional chronologies (a single regional chronology each for RW, EW and LW). Tree-ring series were detrended to remove non-climatic variation associated with aging of the tree and ecological events such as suppression and release within the tree-stand. The spline used to detrend the series had a 50% response wavelength equal to 70% of the length of each ring-width series, retaining an approximate upper limit of variance found at 65 year frequencies. Regional standard (detrended index chronology); [6]) and residual ('prewhitened'; autocorrelation is removed using autoregressive modelling) [26]) chronologies were produced and were subsequently used throughout the analyses.

Table 1. Chronology statistics for each site incorporated into the regional chronology.

Site Name	Code	Chronology Type[a]	# Cores	# Trees	Chronology Interval	Mean Length (Yrs)	Mean Sensitivity	Interseries Correlation	Between tree Correlation	Year EPS >0.85	Year SSS >0.85
Devon Farm	DEV	RW	26	14	1854-2005	83.0	0.291	0.582	0.421	1913	1910
		EW					0.239	0.338	0.181	* never	1920
		LW					0.515	0.555	0.402	1917	1910
Faffard's	FAF	RW	33	19	1886-2006	77.2	0.220	0.573	0.386	1918	1916
		EW					0.219	0.344	0.178	1941	1928
		LW					0.407	0.565	0.335	1920	1917
Hill Side	HIL	RW	14	7	1907-2004	85.9	0.306	0.603	0.486	1908	1908
		EW					0.222	0.369	0.19	* never	1910
		LW					0.517	0.595	0.478	1908	1908
Junction	JCT	RW	14	9	1813-2007	131.0	0.239	0.539	0.37	1891	1857
		EW					0.231	0.254	0.103	* never	1891
		LW					0.548	0.342	0.326	1893	1867
Wildlife Lands	WLL	RW	50	27	1831-2006	100.9	0.251	0.674	0.525	1883	1866
		EW					0.219	0.254	0.247	1902	1892
		LW					0.544	0.625	0.505	1883	1866

[a]Chronology Type: RW - annual ringwidth; EW - earlywood; LW - latewood

2.2 Instrumental climate data

Regional climate variables were derived from gridded climate data for the Prairie region (45° to 60° N and 90° to 120° W). Monthly temperature data were extracted from the CRUTEM3+HadSST2 gridded (0.5°) dataset [27] while monthly precipitation were extracted from Mitchell and Jones' [28] CRU TS 3 - New World gridded (0.5°) dataset. These data were reduced to annual and seasonal (DJF, MAM, JJA, SON) means and totals. Gridded Palmer Drought Severity Index (PDSI) records [29] were also extracted for the Prairie region and evaluated on a month to month basis with the regional chronologies. Because of the scarcity of climate stations on the Canadian Prairies prior to the 20th century [8], all data records were evaluated for the 20th century only (1900 - present).

2.3 PDSI reconstructions

Tree-ring indices were examined for correlations with PDSI, and climate data for the grid point location having the highest absolute correlation (51.25°N, -103.75°W, near Estevan, SK, found with each regional chronology), were extracted for reconstruction purposes. For the regional tree-ring indices used as predictors of PDSI, the length of the chronology was limited to the segment where the subsample signal strength (SSS; a measure of the chronology's ability to retain its original signal back through time; [30]) was above 0.85 (Table 1). Linear regression models were developed by a user-written MATLAB function [31] to estimate PDSI from a set of standardized regional tree-ring predictors through a forward step-wise procedure. Forward and negative lags of up to two years allowed the model to accommodate relationships between drought and tree growth response in the growth year and two years following [6]. The model was validated using the leave-*n*-out cross-validation method. Regression residuals were tested for autocorrelation using the Durbin-Watson test [32], and the mean variance inflation factor (VIF) was calculated to detect multicollinearity in the models.

3. Results and discussion

3.1 Regional chronologies

Standardized and residual regional chronologies (RW, EW and LW), show the coherence of ring-width variability over the entire length of the chronologies (1813 - 2007) exhibiting similar patterns of growth over long-term changes (Figure 2). The chronologies display the same pattern of sustained high growth in the mid 1940s to late 1950s, and suppressed growth in the mid 1800s, 1920s and late 1900s. Based on series intercorrelations and between-tree correlations (Table 1), it would appear that EW chronologies may be less sensitive and responsive to climatic fluctuations influencing tree-ring growth. Given that these two statistical parameters (inter- and between-tree correlations) have similar values for both LW and RW chronologies, it further suggests that latewood widths are more closely related to annual ring-width indices in the timing and magnitude of tree-ring growth response to climatic fluctuations. Correlations among the three chronology parameters (RW, EW, and LW; Figure 3) provides further evidence to support this claim, suggesting that LW indices capture similar, coherent signals, most likely related to regional climate, as depicted in RW series; and could therefore be useful as an as of yet underutilized archive of dendroclimatological material.

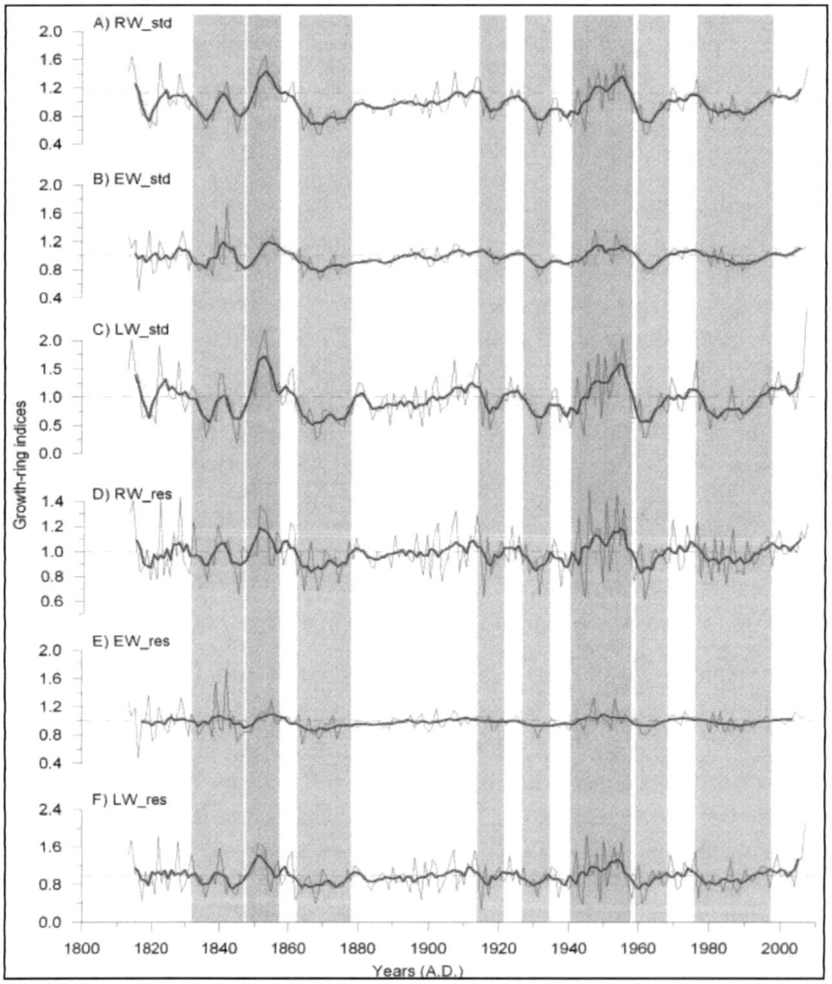

Figure 2. Standardized (std) and residual (res) growth-ring indices for the period 1813-2007. Chronologies are smoothed by a 9 year running average (red). Pink shading indicates years of suppressed growth (below the mean), and blue indicates periods of rapid growth above the mean.

IOP Conf. Series: Earth and Environmental Science **9** (2010) 012016 doi:10.1088/1755-1315/9/1/012016

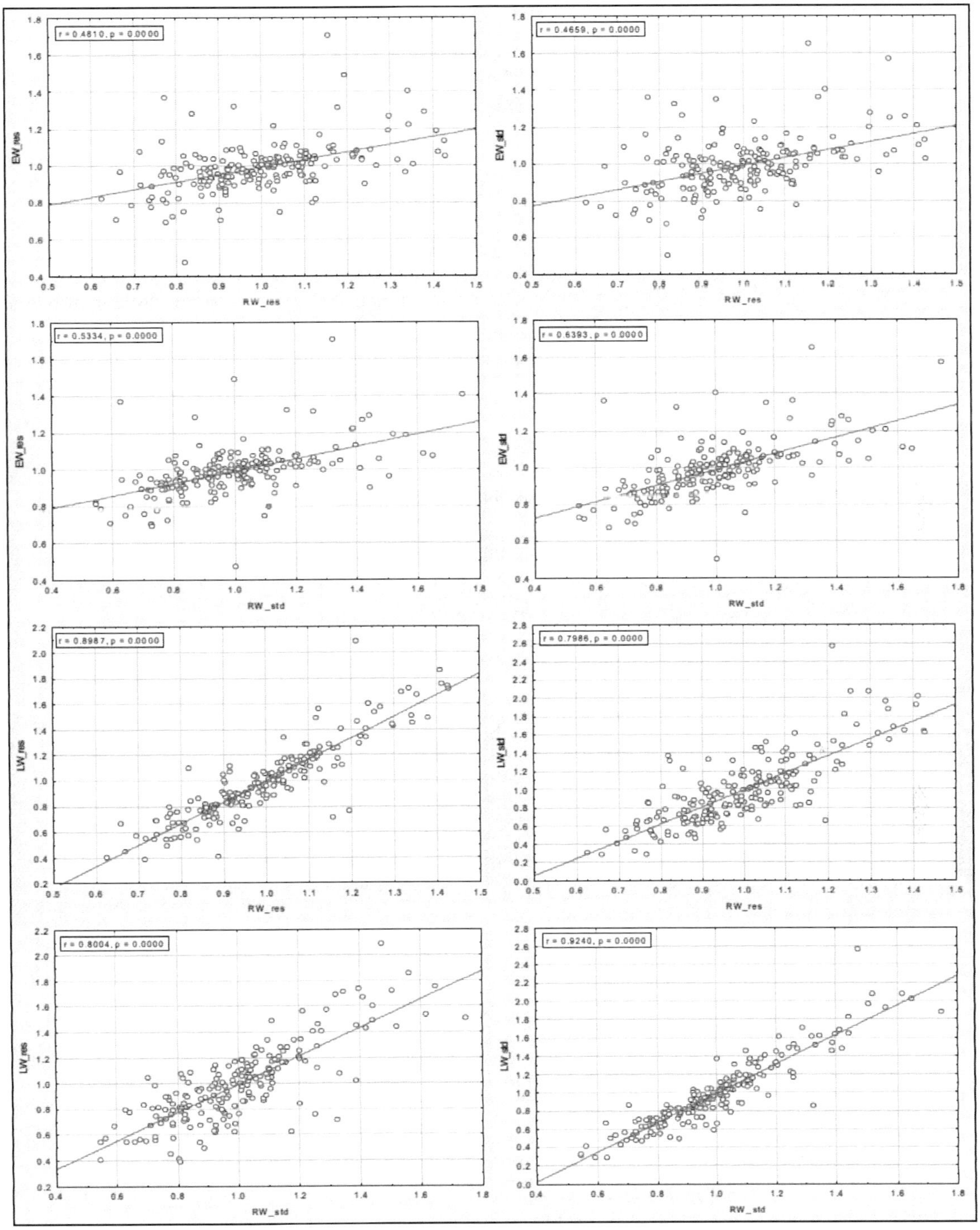

Figure 3: Correlations between annual ringwidth (RW), earlywood (EW) and latewood (LW) chronologies. Correlations are significant at the $p < 0.00001$ level.

3.2 Growth responses to climate

With both standard and residual regional chronologies, RW, EW and LW ring-width indices are positively correlated with precipitation and PDSI values, and negatively correlated with temperature (Figure 4). This is an indication that moisture stress is the major limitation to growth of the species at its western ecological range [8, 13]. Residual indices of each chronology had higher and more widespread correlations with both precipitation and temperature however standardized indices were more strongly correlated with PDSI data values for the Prairie region (Figure 4).

3.2.1 Temperature correlations. In general, RW, EW and LW regional indices have a weak negative correlation with May temperature ($r = -0.3$ to -0.4; $p < 0.05$, results not shown for correlations with standard chronologies; Figure 4) over a relatively wide area, encompassing much of the Prairie region under study. Annual and seasonal temperatures had no significant correlations with either the residual or standardized chronologies (results not shown). The weak negative growth responses suggest that regionally higher spring temperatures impose high evapotranspiration demands, resulting in reduced annual ring growth.

Earlywood chronologies had the most significant and widespread negative correlations to temperature, corresponding to the negative effect of temperatures earlier in the year. During the cold (winter) months, ring porous species such as bur oak enter a quiescence phase in which it gains the ability to respond to growth-promoting conditions. During quiescence, before photosynthates are yet to be produced, the tree uses stored reserves from the previous growing season [33], and only changes at the ultrastructural level occur within a tree (*i.e.* EW vessel formation; [34]). Visible signs of activity appear when the conditions become favourable for LW growth; thus, warming spring temperatures and increasing evapotranspiration demands, correlate negatively with the production of EW growth (Figure 4), potentially deeming it a proxy for spring temperatures.

3.2.2 Precipitation correlations. Annual, EW and LW growth indices are primarily correlated with summer (JJA) precipitation ($r = 0.2 - 0.5$; $p < 0.05$, results not shown for correlations with standard chronologies; Figure 4). Correlation analyses show that regional RW and LW display a similar sensitivity to summer precipitation, while the moisture signal in EW is much weaker, further supporting the closer relation of LW to RW widths than to EW widths. Correlations between spring, autumn, winter (prior autumn and winter) and all ring-width series were not significant at the regional level (results not shown). Although chronologies were found to be significantly correlated with summer precipitation only, residual growth indices also correlate with annual (June to May) precipitation (not shown), "but this correspondence likely reflects the importance of summer rainfall on the Prairies rather than a real physical relationship between ring-width and total annual precipitation" [8]. If an 'actual' significant annual correlation were to exist, growth indices would likely have some correlation with other seasons, most likely with prior winter precipitation, reflecting the importance of winter snowfall to soil moisture recharge and growth resumption in the spring [13, 35].

3.2.3 PDSI correlations and reconstructions. Unlike analysis of temperature and precipitation, standardized RW, EW and LW regional chronologies have stronger positive correlations with monthly PDSI values than do residual chronologies. PDSI data and regional chronology correlations were significant for all months (January through December; results not shown), however the strongest association was found during May ($r = 0.4\text{-}0.6$; $p < 0.05$, results shown for correlations with standard chronologies only; Figure 4). Correlation throughout the year results from the autocorrelation with both standardized chronologies and the PDSI. Successive years of moisture limited growth, will result in suppressed growth in the following years. Similarly, PDSI for a given month or season depends partly on previous months' moisture conditions through soil-moisture storage and autoregressive terms in the computation algorithm [36]. Therefore, monthly PDSI values reflect moisture conditions not

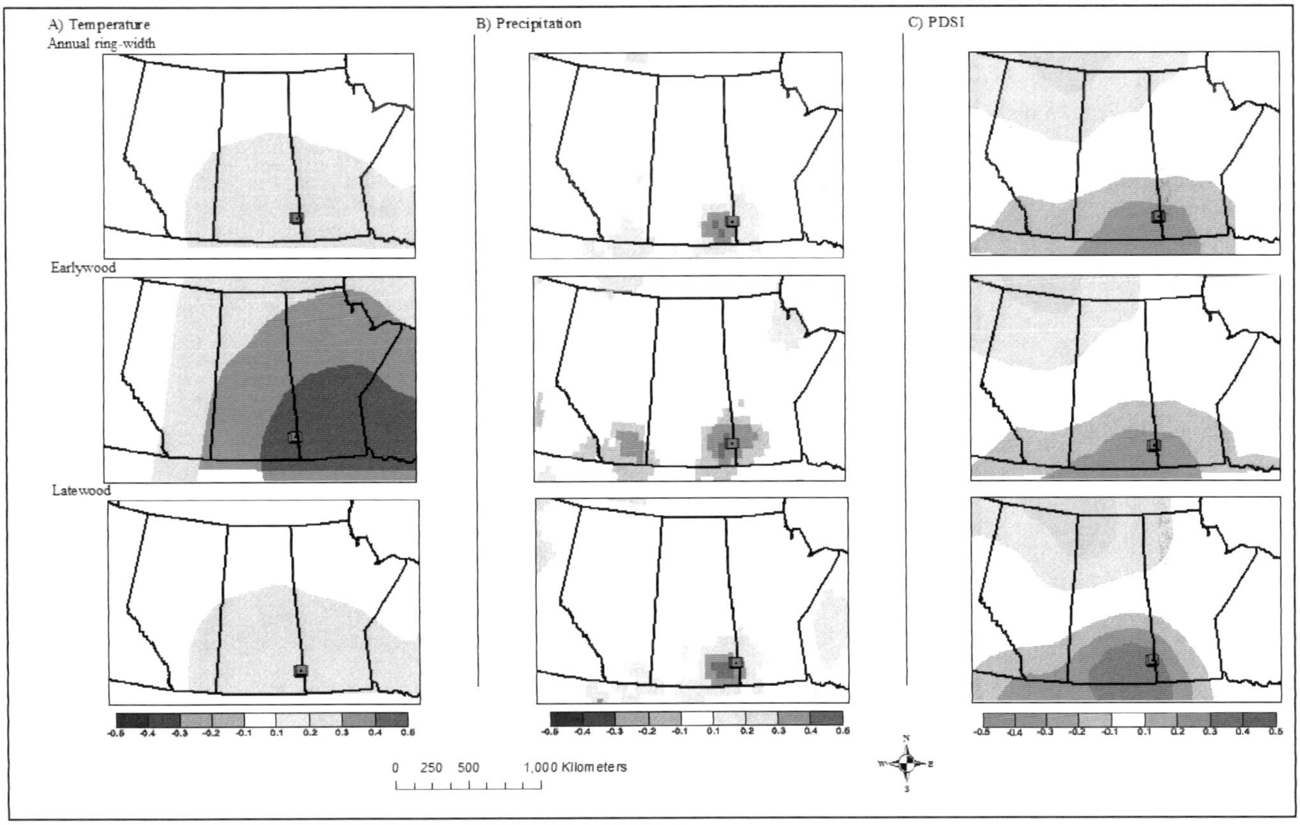

Figure 4. Correlation analysis between A) monthly (May) mean temperature; B) summer (JJA) precipitation; C) monthly (May) PDSI, for the period 1900 - 2003. Correlations shown are significant at $\alpha = 0.05$. Site locations are denoted by the green squares.

only for that month, but also several months preceding the timeframe in question [19]; which is ultimately reflected by the autocorrelated growth responses of the tree-ring widths.

Latewood chronologies again display more similarities in levels of significance and spatial coverage to the RW indices. Because of the strong correlations between standard chronologies and PDSI data, reconstruction models (Figure 5; Table 2) were generated for RW and LW to assess the applicability of these indices as proxies of PDSI. Regression models were skilful and significant (Table 2) and the Durbin-Watson statistic showed that residuals from both models were uncorrelated [37]. In the verification period, reconstructions were significant and the positive values for the reduction of error (RE) statistic indicate that the models have significant predictive capabilities [6].

Both models accounted well for the timing and magnitude of severe 20th century summer droughts (Figure 5) in the region, and were successful in depicting the peaks and lows of interannual variations in the instrumental data; PDSI values ranging from 0 to -4 (0 being no drought, -4 representing severe droughts). Although both models are able to adequately reconstruct PDSI, the LW model was able to account for more variance than when using the RW indices as predictors of PDSI (35.5% and 32.5%, respectively). The mean variance inflation factor (VIF; Table 2) is below the critical threshold for reconstruction purposes [38], suggesting that multicollinearity is not an issue, and the model is valid. The use of LW widths as proxies allows for sub-annual analysis and a more in depth understanding of seasonal hydroclimatic variability for the Prairie region; an area that frequently is faced with the consequences of drought and heavy reliance on the summer growing season for the sustainability of its agricultural economy.

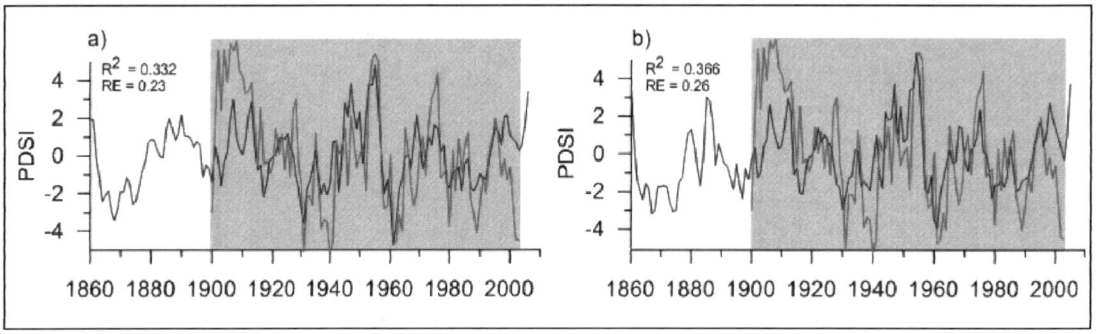

Figure 5. Total annual ringwidth (a) and latewood (b) proxy reconstructions of PDSI near Estevan, SK for the period 1860 to 2005. Red lines are observed instrumental data values, and blue lines are reconstructed values of PDSI. Shading of the blue box indicates the calibration period for the models (1900 - 2003).

Table 2. Regression statistics of PDSI reconstructions.

Name	Period	Length (yrs)	Calibration Period	Lag	Predictors	R^2	R^2_{adj}	RE	SE	DW	VIF_{max}
LW_PDSI	1860-2005	146	1900-2003	±2	2	0.366	0.353	0.26	2.27	H_0	1.18
RW_PDSI	1860-2005	146	1900-2003	±2	2	0.332	0.325	0.23	2.45	H_0	1.18

Proxy	Model
LW	$-6.0600 + 4.5715*LW + 1.7044*LW_{+1}$
RW	$-10.5025 + 6.0773*RW + 4.6406*RW_{+1}$

R^2: multiple correlation coefficient.

R^2_{adj}: Multiple correlation coefficient adjusted for degrees of freedom.

RE: Reduction of Error statistic; any value greater than 0 indicates value in the reconstruction.

SE: Standard deviation of the differences between the actual values of the dependent variables and the predicted values

DW: Durbin-Watson statistic: H_0 - no first order autocorrelation in residuals.

VIF_{max}: Variance inflation factor, used to detect multicollinearity in the models.

4. Conclusions

This study demonstrates that *Q. macrocarpa* can provide multiple dendroclimatic proxies at annual and sub-annual time scales, for investigating large scale climatic fluctuations that influence the climate of the Prairie region. Results indicate that RW, EW and LW chronologies from southeastern Saskatchewan capture signals related to regional moisture and drought conditions. Significant negative correlations to temperature and positive associations with precipitation and PDSI indicate that trees in this region are thus limited by moisture availability and are sensitive to drought. Correlations between RW, EW and LW chronologies suggest that LW widths are more representative of annual ring-widths, and are the best proxy of seasonal fluctuations in climatic data, explaining more variance in PDSI.

Reconstructions indicate that regional LW widths are the most significant ring-width indices as predictors for examining Prairie drought. Although the difference in amount of variance explained by the LW and RW reconstructions is minimal, it does provide evidence that LW can be used as a proxy for climate variables. Adjusting for the inter-correlation with earlywood-width may prove useful as an improvement to increase the signal retained in the LW chronologies [19], and should be investigated further.

This preliminary study using LW as a proxy for climate variability suggests that the geographical expansion of coverage by LW chronologies and further development of statistical methods may lead to successful reconstructions of indices of synoptic-scale drought, other than PDSI, such as the El Niño/Southern Oscillation (ENSO) and Pacific Decadal Oscillation (PDO), patterns that influence the climate of the Prairie region. Further research to establish a network of *Q. macrocarpa* chronologies in this region of few dendroclimatic records, will contribute to a better understanding of the regional climate variability.

Acknowledgments

This research was supported by the Natural Sciences and Engineering Research Council of Canada. For field and lab assistance we thank Cesar Perez-Valdivia, Michael Felgate, Jeremy Pittman, Mariya Krestyanska, Suzan Lapp, Jonathan Barichivich, Jill Dallamore-Deschamps, and Svetlana Esenkulova. Helpful and constructive comments were provided by Scott St. George and an anonymous reviewer.

References

[1] Kissinger M and Rees W E 2009 *Ecol. Econ.* **68** 2309-2315
[2] McGinn S M and Shepherd A 2003 *Can. J. Soil Sc.* **83** 623-630
[3] Intergovernmental Panel on Climate Change 2001 *Climate change 2001: The scientific basis. Summary for policymakers and technical summary of the Working Group I Report (Third Assessment)*
[4] McGinn S M and Shepherd A 2003 *Can J Soil Sci* **83** 623-630
[5] St George S, Anderson T W, Forbes D, Lewis C F M, Nielsen E and Thorleifson L H 2001 *Climatic extremes in southern Manitoba during the past millennium* (Final Report to the Canadian Climate Change Action Fund) Environment Canada
[6] Fritts H C 1976 *Tree Rings and Climate* (New York: Academic Press)
[7] Luckman B D 1990 *Mtn Res Dev* **10** 183-195
[8] St. George S, Meko D M, Girardin M-P, MacDonald G M, Nielsen E, Pederson G T, Sauchyn D J, Tardif J C and Watson E 2009 *J. Clim.* **22** 689-710
[9] Sauchyn D J and Skinner W R 2001*Can. Water Res. J.* **26** 253-273
[10] Lawson M P 1980 *Tree-Ring Bull.* **40** 1-11
[11] Stockton C W and Meko DM 1983 *J. Clim. App. Met.* **22** 17-29
[12] Blasing J T and Duvick D N 1984 *Nature* **307** 143-145
[13] Case R A 2000 *Dendrochronological investigations of precipitation and streamflow for the Canadian Prairies. (PhD dissertation)* (Los Angeles: University of California)
[14] Woodcock D W 1989 *Can. J. For. Res.* **19** 639-644
[15] Panshin A J and de Zeeuw C 1970 *Textbook of Wood Technology* (New York: McGraw-Hill)
[16] Drunasky N and Struve D K 2005 *Urb For Urb Green* **4** 13-22
[17] St George S, Nielsen E, Conciatori F and Tardif J 2002 *Tree-Ring Res* **58** 3-10
[18] Carlquist S 1975 *Ecological Strategies of Xylem Evolution* (Berkley: University of California)
[19] Meko D and Baisan C H 2001 *Int. J. Clim.* **21** 697-708
[20] Conkey L E 1979 *Tree-Ring Bull* **39** 29-38
[21] Brunstein F C 1996 *Arc Alp Res* **28** 65-76
[22] Roden J 2008 *Chem Geol* **252** 72-79
[23] Stokes M A and Smiley T L 1968 *An Introduction to Tree-ring Dating* (Tuscon: University Arizona Press)
[24] Holmes R L 1983 *Tree-Ring Bull.* **43** 69-78
[25] Grissino-Mayer H D 2001 *Tree-Ring Res.* **52** 205-221
[26] Cook E R 1985 *A time series approach to tree-ring standardization. (PhD dissertation)* (Tuscon: The University of Arizona)
[27] Ranier N A,Brohan P, Parker D E, Folland C K, Kennedy J J, Vanicek M, Ansell T and Tett S F B 2006 *J. Clim.* **19** 446-469
[28] Mitchell T D and Jones P D 2005 *Int. J. Clim.* **25** 692-712
[29] Dai A, Trenberth K E and Qian T 2004 *J. Hydromet.* **5** 1117-1130
[30] Wigley T M, Briffa K R and Jones P D 1984 *J. Clim. Appl. Meteorol.* **23** 201-213.
[31] Meko D 2005 *GEOS 585A, Applied Time Series Analysis* (Arizona: Laboratory of Tree Ring Research)
[32] Ostrom C W Jr 1990 *Time Series Analysis: Regression Techniques, Quant. Appl. Soc. Sci.* vol 07-009, 2nd ed. (Newbury Park, California: Sage)
[33] Speer J H 2008 *Fundamentals of Tree-Ring Research (In review)*
[34] Farrar J J and Evert R F 1997 *Trees* **11** 191-202
[35] Robertson P A 1992 *Amer Mid Nat* **128** 218-326
[36] Palmer W C 1965 *Meteorological drought* (Research Paper No. 45) (US Weather Bureau, Department of Commerce)

[37] Draper N R and Smith H 1981 *Applied Regression Analysis*, 2nd ed (New York: John Wiley)
[38] Haan C T 2002 *Statistical Methods in Hydrology,* 2nd ed (Ames, Iowa: Iowa State University Press)

IOP Conf. Series: Earth and Environmental Science **9** (2010) 012017 doi:10.1088/1755-1315/9/1/012017

Teberda Valley runoff variability (AD 1850–2005) based on tree-ring reconstruction (Northern Caucasus, Russia)

V V Matskovsky[1], E A Dolgova[1] and O N Solomina[1]

[1] Institute of Geography, Russian Academy of Sciences, 29 Staromonetniy pereulok, Moscow, Russia

E-mail: matskovsky@gmail.com

Abstract. Seven tree-ring chronologies are used to reconstruct Teberda River (Northern Caucasus, Russia) smoothed runoff for May, July and August. Six chronologies were developed from *Pinus sylvestris* and one from *Abies nordmanniana.* Tree growth is significantly, but weakly, correlated with maximum temperatures (negatively) and relative humidity (positively) during summer. All possible combinations of seven predictors were tried to get the best result on the cross-validation. Two of three reconstructions showed high wavelet coherence with instrumental data on decadal timescales and were analysed for spectrum stability. Minima of moving trends at the end of the reconstructions along with weakening of decadal cycles may be a marker of significant change of Teberda River hydrological regime during the second half of the 20th century.

1. Introduction

Teberda River is a tributary of Kuban' (Azov Sea basin), the largest river of Krasnodarskiy Kray, an important agricultural region of the Russian South. Teberda is 60 km long with the watershed surface equal to 1080 km^2, the mean watershed altitude is 2210 m. 55.8% of Teberda River's runoff (measured at Teberda hydrological station) is provided by snow and ice melt. 60% of runoff occurs in summer, 17% in the fall, 5% in winter and 18% in spring. Mean annual water discharge is 26.8 m^3/sec [1]. Teberda monthly runoff correlates with spring temperature (up to 0.62 for April) and with autumn precipitation (up to 0.72 for October). Significant negative trends were identified both in summer and full year runoff for AD 1927-2005. Extending instrumental observations into the past is essential for understanding of the runoff long-term variability and response to changing climate.

In this study we use seven tree-ring chronologies developed from two species to reconstruct streamflow characteristics prior to the onset of instrumental monitoring. Principal component regression and wavelet analyses are used to extend runoff data into the past, to estimate the quality and informational content of the reconstructions and to make inferences about the extremeness of the measured data variability.

2. Materials and methods

2.1. Tree-ring data

The study area is located in the Caucasus Mts. in the territory of Teberda Biosphere Reserve and Elbrus National Park. The current timberline of Scots pines (*Pinus sylvestris L.*) rises up to the elevation of 2500-2700 m, depending on the slope orientation and local environmental conditions. Tree ring samples (cores) were collected from nine sites in the vicinity of the upper treeline (from 2200 to 2500 m a.s.l.) and at the glacier's forefields. Most of the sites are located on south-facing slopes and have poor soils. Location of these sites are shown on the figure 1.

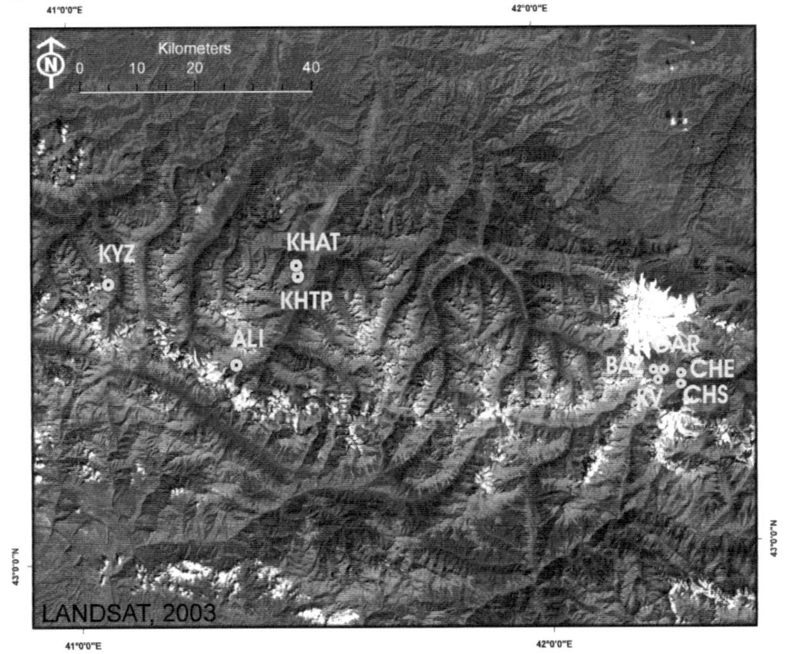

Figure 1. Tree-ring sites (yellow) and Teberda River valley (green). Mt. Elbrus (white) is clearly seen at the right side of the figure.

Living trees of Scots pine (*Pinus sylvestris L.)* and fir (*Abies nordmanniana (Stev.) Spach*) were sampled according to standard dendroclimatological principles and conventions [2], [3]. LINTAB equipment was used for measuring annual ring-width with a precision of 0.01 mm. To cross-date tree-ring series and compare them graphically, we used Rinntech TSAP 0.53 software. To cross-date and check for missing rings or dating errors we used COFECHA software [4]. Samples that showed low correlation (<0.33) with the master-chronology were excluded from the analysis. The final dataset comprised 145 tree-ring series.

We used program ARSTAN software [5] for developing the chronologies. Tree-ring series were standardized using a conservative procedure to maintain in the series as much low-frequency variance as possible. Negative exponential or linear curves were used for removing the non-climatic signals. All the chronologies were detrended by ratio method. The ability of each chronology to represent the ideal population signal was assessed using the mean between-tree correlation and the expressed population signal (EPS) statistics [6].

2.2. Runoff data

Monthly values of Teberda River runoff at Teberda hydrological station for AD 1927-2005, corrected for nonstationarity (that emerged from different measurement techniques) were provided by Teberda Biosphere Reserve. The data contain missing values from August of 1941 to December of 1947 and from January of 1971 to January of 1972 (missed 102 of 948 months in total). Missing values of instrumental runoff data were ignored in all the analyses. There have not been any dams or diversions that could affect runoff values during the instrumental period.

2.3. Methods

We built a principal component chronology to extract a common climatic signal from our local chronologies. This component chronology showed a strong response to climatic variability (calculations in Dendroclim2002 [7]): it correlated negatively with May-July maximum temperature and positively with May-July relative humidity. A significant correlation (at 99.9% confidence level) was identified between PC chronology indices and July Teberda River runoff (r=0.41). But none of the chronologies showed a really strong response to any individual climatic variable. However, the absence of a strong linear correlation with a single climate variable does not mean that tree-growth in this region is not sensitive to climate [8]. Both tree rings and streamflow are integrators of several factors: temperature, precipitation and soil parameters. It allows the conclusion that correlation between them is not accidental.

We used principle component regression [9] to generate the runoff reconstruction models, separately for each of the 12 months of the year and for the full year. Seven standard chronologies listed below were used as predictors. There are also significant correlations of tree rings and runoff at different lags. We also tried to include lagged tree ring chronologies as predictors in the regression model but it didn't improve the reconstructions. To determine the best set of predictors, all possible combinations were tried and the set that showed maximum R^2 statistic on leave-one-out cross-validation was chosen.

Wavelet coherence analysis [10] was applied to determine common properties of instrumental and reconstructed data. Continuous wavelet transform with the Morlet base wavelet [11] was used for analysis of spectral characteristics of reconstructions through time.

The following technique was used to test for significance of the trends. Firstly the time series was detrended and the AR(1) process was estimated from the residuals. Then Monte Carlo simulation was used to generate 10000 independent time series from the estimated process and a trend was calculated for each of the generated series. Finally the 2.5 and 97.5 percentiles of these trends' distribution were taken as lower and upper bounds for testing the significance of the trend.

3. Results

3.1. Chronologies

We developed six pine (KHTP, KV, KYZ, CHS, GAZ and BAZ) and one fir (ALI) ring-width chronologies. The KHTP chronology consists of samples from sites KHTP and KHAT, and the CHS chronology consists of samples from sites CHS and CHE (figure 1). Commonly used statistics such as standard deviation, first order autocorrelation, mean sensitivity – a measure of the relative difference in width between consecutive rings [2] – are listed in table 1. All the chronologies are characterized by relatively low mean sensitivity and high autocorrelation.

Table 1. Tree-ring chronologies statistics.

Site name	KYZ	KV	KHTP	GAR	CHS	BAZ	ALI
First year	1550	1640	1678	1693	1738	1660	1800
Last year	2006	2002	2005	2002	2002	2004	2005
Cores (trees)	20 (10)	8 (4)	57 (29)	13 (7)	15 (8)	18 (11)	24 (13)
Inter-series correlation	0.649	0.551	0.507	0.614	0.526	0.436	0.282
Standard deviation	0.221	0.176	0.221	0.221	0148	0.140	0.125
Mean sensitivity	0.157	0.125	0.155	0.128	0.110	0.119	0.118
1st order autocorrelation	0.648	0.550	0.505	0.612	0.524	0.435	0.282
Mean segment length	227	283	175	160	160	233	114
EPS > 0.8	1797	1863	1752	1939	1833	1768	1850

3.2. Reconstructions

Annual reconstructions showed unreliable results on cross-validation. Consequently we smoothed reconstructed series with a 10-year moving average filter. Three of thirteen smoothed reconstructions (for each month and for the full year) showed R^2 higher than 0.6 in cross-validation: May runoff (the best set of predictors is ALI, BAZ, CHS and KHAT, R^2=0.6487), July runoff (ALI, BAZ and KHAT, R^2=0.6567) and August runoff (ALI, CHS and KHAT, R^2=0.6615) (figure 2a). During these months occurs 12%, 22% and 17% of annual runoff respectively (according to instrumental records). The three months combined account for more than half the annual runoff. We considered only those parts of the reconstructions for which EPS of all chronologies-predictors is higher than 0.8. In all the cases the ALI chronology limited the start to AD 1850.

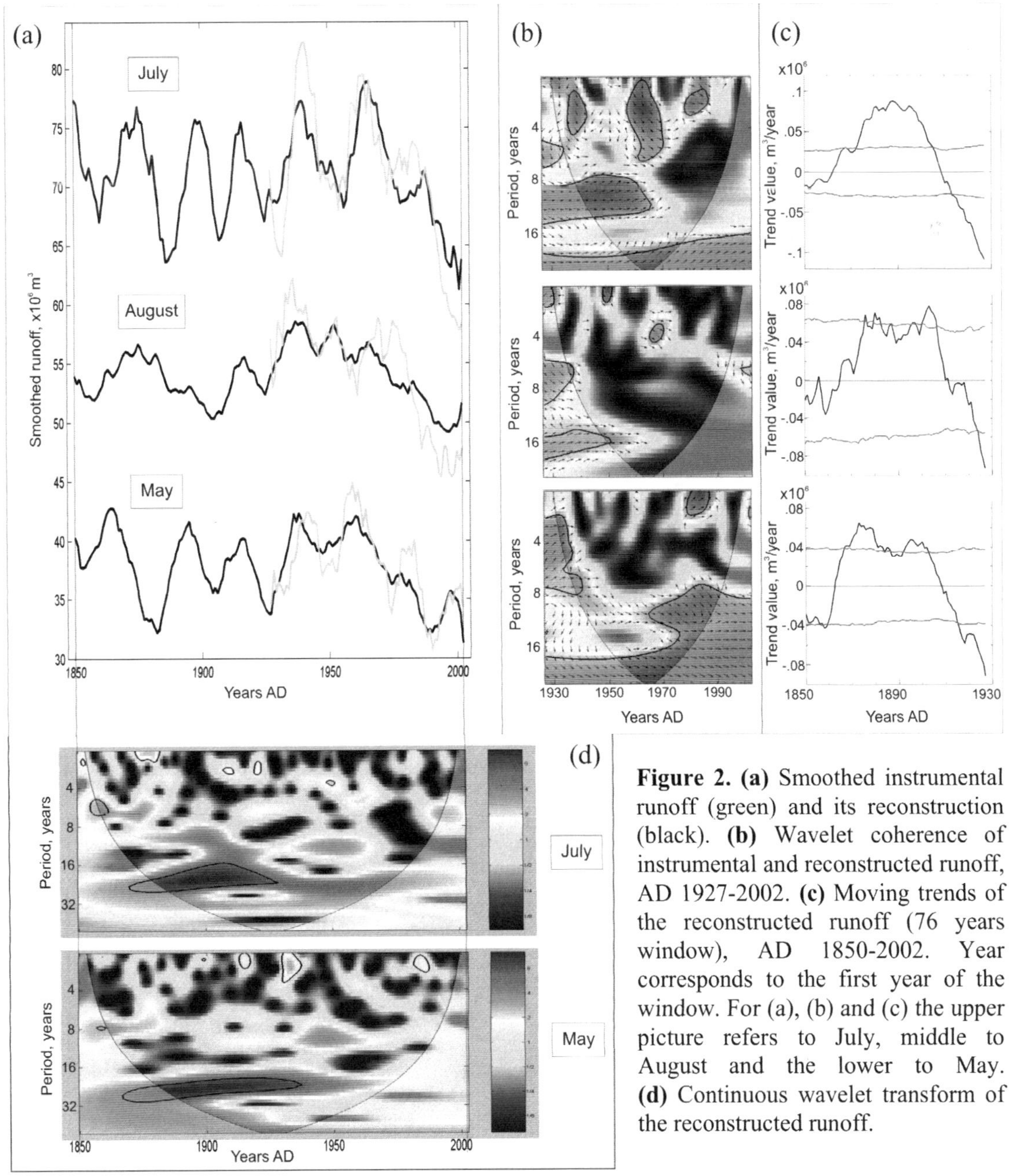

Figure 2. (a) Smoothed instrumental runoff (green) and its reconstruction (black). **(b)** Wavelet coherence of instrumental and reconstructed runoff, AD 1927-2002. **(c)** Moving trends of the reconstructed runoff (76 years window), AD 1850-2002. Year corresponds to the first year of the window. For (a), (b) and (c) the upper picture refers to July, middle to August and the lower to May. **(d)** Continuous wavelet transform of the reconstructed runoff.

Two of three reconstructions and the sum of the three reconstructed months reach their minimum values in 1997 (ignoring last 5 years because of 10-years smoothing). May's reconstruction has the second lowest value in 1990.

Wavelet coherence analysis was applied to instrumental and reconstructed (not smoothed) data to estimate the ability of reconstructions to reproduce low-frequency variations. May and July reconstructions showed strong and stable coherence with instrumental records on periods more than 25 years, while August reconstruction did not (figure 2b).

Changes of decadal variability of May and July are obvious from our smoothed reconstructions. Wavelet analysis can define exact year of such changes. Continuous wavelet transforms for runoff reconstructions (not smoothed) (figure 2d) show weakening of low frequency cycles (16 to 32 years periods) after AD 1929 for July and after AD 1938 for May. We didn't perform wavelet analysis for the August reconstruction because of weak wavelet coherence of reconstructed and instrumental data. For wavelet spectra diagrams, watery colors show the cone of influence, and the black border shows regions of greater than 95% confidence for a red-noise process.

3.3. Trends

Runoff of all three months (May, July and August) as well as full year's runoff have significant negative trends for the period AD 1927-2005, with values -0.063, -0.140, -0.165 and -0.429 $x10^6\ m^3$/year correspondingly.

We calculated moving trends for our three reconstructions (without smoothing) and for their sum for AD 1850-2002 in 76 years moving window. The years on the plots correspond to the left side of the window, namely 1850 means years from 1850 to 1925 (figure 2c). All trends showed their minimum values for the last window (AD 1927-2002). Trends for other moving windows (60, 50 and 40 years, not shown) sometimes have minimum values in another (not last) windows. The sum of the reconstructions showed its minimum in the last window in all the cases.

4. Discussion and summary

Although we didn't create annually resolved quantitative reconstructions of Teberda River runoff because of insufficient correlation between tree growth and hydrological parameters, our smoothed reconstruction of May, July and August runoff can tell much about low-frequency variations of these parameters.

The multidecadal variability in the 15-30-year band apparent for May and July runoff become insignificant after AD 1930-40. The genesis of this variability is poorly understood but its weakening can result from significant environmental changes. Almost all explored characteristics, such as trends, spectra and smoothed values of the series have minima near the end of the reconstructions. These extremes along with weakening of decadal cycles may be a marker of significant change of Teberda River hydrological regime.

We made our reconstructions only back to AD 1850, yet some of our chronologies begin in 16th century. Adding new samples to these chronologies will increase the common signal and let us extend our reconstructions into the past.

Acknowledgments

We thank V. Mikhalenko, V. Gryaznova and others who helped with the collection and analysis of the tree-ring data, A. Grinsted for his Crosswavelet and Wavelet coherence analysis Matlab package, S. Rupper and two anonymous reviewers for reviewing this manuscript prior to publication. This research is supported by RFBR grants 07-05-00410 and 09-05-10043. V. Matskovski thanks PAGES and the U.S. National Science Foundation for their financial support to attend the 1st PAGES Young Scientitsts Meeting and 3rd Open Science Meeting.

References

[1] Lurye P M, Panov V D and Tkachenko Yu Yu 2005 *Kuban' River: hydrography and runoff regime* (in russian) (Sankt-Petersburg: Hydrometeoizdat)
[2] Fritts H C 1976 *Tree Rings and Climate* (London: Academic Press)

[3] Cook E R and Kairiukstis L A 1990 *Methods of Dendrochronology: applications in the environmental sciences* (Dordrecht: Kluwer)

[4] Holmes R L 1983 Computer-assisted quality control in tree-ring dating and measurement *Tree-Ring Bulletin* **43** 69-78

[5] Cook E R and Holmes R L 1984 *Program ARSTAN and Users Manual* (NY: Lamont Doherty Geological Observatory)

[6] Wigley T M L, Briffa K R and Jones P D 1984 On the average value of correlated time series, with applications in dendroclimatology and hydrometeorology. *J. Climate Appl. Meteor.* **23** 201–13

[7] Biondi F and Waikul K 2004 DENDROCLIM2002: A C++ program for statistical calibration of climate signals in tree-ring chronologies *Computers & Geosciences* **30** 303–11

[8] St. George S, Meko D M and Evans M N 2008 Regional tree growth and inferred summer climate in the Winnipeg River basin, Canada, since AD 1783 *Quat. Res.* **70** 158–72

[9] Aivazyan S A, Yenyukov I S and Meshalkin L D 1985 *Applied statistics. Study of relationships* (in russian) (Moscow: Finansy i Statistika)

[10] Grinsted A, Moore J and Jevrejeva S 2004 Application of the cross wavelet transform and wavelet coherence to geophysical time series *Nonlinear Processes in Geophysics* **11 5/6** 561-6

[11] Torrence C and Compo G P 1998 A practical guide to wavelet analysis *Bull. Amer. Met. Soc.* **79 1** 61–78

Mg/Ca variability of the planktonic foraminifera *G. ruber s.s.* and *N. dutertrei* from shallow and deep cores determined by electron microprobe image mapping

J Fehrenbacher, P Martin

University of Chicago, Department of Geophysical Sciences, Chicago, IL 60637

E-mail: jsf1@uchicago.edu

Abstract. The Mg/Ca ratio of planktonic foraminifera is widely used for reconstructing sea surface and thermocline temperatures. Dissolution, driven by changes in the bottom water calcite saturation state (a function of water mass chemistry and depth/pressure), lowers the bulk Mg/Ca ratio which results in an implied cooler reconstructed temperature. Here we employ electron microprobe analysis (EMPA) to characterize how dissolution alters shell Mg/Ca variability. Using EMPA we have generated Mg/Ca image maps of two species widely used in paleoceanographic reconstructions: *Globogerinoides ruber sensu stricto* (*G. ruber s.s.*) and *Neogloboquadrina dutertrei (N. dutertrei)*. We first characterize the Mg/Ca ratio in the better-preserved shells from a shallow core (2.8 km) and compare those characteristics to the samples from a deep core (4.0 km) obtained from the same locality (Ceara Rise, western equatorial Atlantic). Though preliminary, results suggest EMPA Mg/Ca ratio mapping can reveal characteristics of well preserved shells that can be used to assess changes in the Mg/Ca ratio that occur with dissolution.

1. Introduction

The Mg/Ca ratio in foraminiferal calcite is an established and widely used proxy for reconstructing sea surface temperature (SST). The exponential temperature-Mg/Ca relationship was first established in inorganic calcite precipitation experiments [1, 2]. The temperature-Mg/Ca relationship in foraminifera was later documented in culturing experiments [3-6], sediment-trap studies [7, 8] and core-top calibration studies [9]. The Mg/Ca ratio increases by 8-10% for most planktonic foraminifera per 1°C increase in temperature [3, 4, 10] although the temperature sensitivity appears to be lower (5-8% per 1°C) in some species that calcify in the thermocline [see for example 11, 12]. Many studies have used the Mg/Ca-SST relationship to reconstruct both sea surface and water column temperatures [13-17].

Shell dissolution lowers the bulk Mg/Ca ratio, which if left unaccounted for, results in an underestimation of past temperatures [12, 18-20]. Dissolution can occur in the water column, on the sea floor, and in the sediment, although most of the dissolution is thought to occur on the seafloor prior to burial [21, 22]. In an effort to minimize the dissolution effect and obtain accurate paleotemperatures, many researchers use samples obtained from cores well above the lysocline or apply a dissolution correction parameter such as water depth or size normalized shell-weight [12, 17, 23]. We recently used the decrease in the Mg/Ca ratio with depth to quantify changes in bottom water calcite saturation state during the Last Glacial Maximum in the tropical Atlantic and Pacific oceans [24-26]. Though it is widely known that dissolution lowers the bulk ratio, the mechanism by which dissolution alters the shell Mg/Ca ratio is poorly understood.

Several recent studies have shown the utility of assessing shell Mg/Ca variability using an electron microprobe, laser ablation ICPMS, and nano-SIMS [27-30]. These studies revealed high variability in the Mg/Ca ratio among different species and within individual tests. For example, several species contain bands of alternating high- and low-Mg calcite. These studies also demonstrated the utility of Mg/Ca imaging as a means of determining inter- and intratest Mg/Ca variability and highlight the need for further investigation for the differences in the Mg/Ca ratios among the different species and with changes in preservation.

In this study we employ the electron microprobe imaging technique to characterize how dissolution alters the distribution of the Mg/Ca ratio in *G. ruber s.s.* and *N. dutertrei*, two species commonly used in paleoceanographic studies. We use samples from a shallow core to characterize the distribution of Mg/Ca within the better-preserved samples and samples from a deeper core to assess changes in shell Mg/Ca variability attributed to dissolution. The core sites used in this study are closely spaced and contain foraminifera that likely calcified under the same surface conditions. The shells settled on the ocean floor at different depths where they experienced different intensities of dissolution due to the bottom water calcite saturation state, ΔCO_3^{2-}; where $\Delta CO_3^{2-} = [CO_3^{2-}]_{in\ situ} - [CO_3^{2-}]_{sat}$. $[CO_3^{2-}]_{in\ situ}$ is a water mass property and $[CO_3^{2-}]_{sat}$ is a function of temperature and pressure. The cores used in this study are bathed in the same water mass, North Atlantic Deep Water (NADW), thus their *in situ* carbonate ion concentrations are similar and the dissolution effect at these locations is driven by pressure (i.e. increased water depth).

2. Samples and Methods

Samples were obtained from core-top intervals from a shallow (KNR110-82; 2.8 km, 4.3°N 43.3°W) and deep (KNR110-50; 4.0 km, 4.9°N 43.2°W) core from the Ceara Rise in the western tropical Atlantic. The core sites are closely spaced (~65 km apart) and thus contain foraminifera that calcified in similar sea surface conditions (annual SST 27.4°C; seasonal SST variability <1°C) [31]. *G. ruber s.s.* and *N. dutertrei* shells were picked from the 250-355μm size fraction. Shells were rinsed in n-pure water and methanol and briefly ultrasonicated to remove fine adhering clays.

Samples were embedded in Araldite epoxy then sanded and polished to reveal cross-sections of the shell wall. For each species, between 8 and 15 shells were analyzed from each sample. When possible, the final chamber and one previously formed chamber (i.e. F-1) was analyzed in each shell. In some samples, due to orientation when preparing the sample, it was not possible to identify the chamber number. The Mg/Ca image maps were generated on a Cameca SX50 electron microprobe operating at a 15 kV accelerating voltage and a beam current of 15nA. Column conditions were chosen to minimize excitation volume. Maps were generated using a 1μm step size and 1s dwell time. Ca and Mg intensities were measured on two spectrometers (PET and TAP respectively) for improved statistics. To take into account 'edge effects' measurements with raw Ca intensities less than 3000 counts per second were disregarded. Raw Mg and Ca intensities were calibrated using a natural dolomite standard [32]. Dolomite standard data was collected before and after each scan. The standard data had a daily standard deviation of ~0.3%. The background for Mg was established by counting for Mg on a Mg-free calcite spar. A 'foraminifera standard' was used to assess reproducibility. For the foraminifera standard, image scans were obtained on the same shell at least once during each week of data collection; Mg/Ca maps generated on the same chamber within the foraminifera standard were reproducible to within ~5%. We used a propagation of errors method to determine the error on the image maps and estimate the error to be ~10% (based on the count rate error and the standard deviation of the standard data). The average Mg/Ca ratios of the individual maps were compared to bulk Mg/Ca measurements made using HR-ICPMS [24].

3. Results

3.1. G. ruber results

Most of the well-preserved *G. ruber* specimens (7 of 8) contain bands of high and low Mg-calcite as reported in other recent studies (Figure 1A and 1B) [29]. The Mg/Ca ratio is highly variable from sample to sample ranging from 1.0 to 11 mmol/mol for individual pixels with average Mg/Ca values for individual image maps ranging from 3.0 – 5.4 mmol/mol with an average of 3.8 mmol/mol (N = 8). The average Mg/Ca values are in good agreement with bulk Mg/Ca values (4.4 mmol/mol) obtained from HR-ICPMS analysis [24]. The higher Mg/Ca values tend to occur in the thin-banded layers, on the inner and outer surfaces of the shells or in the area where two chambers join. The high values on the shell surfaces could be a real feature of the calcification or may be due to Mg/Ca contaminant phases (see for example, Pena et al 2008 [33]).

Similar to the shallow core results, the Mg/Ca ratio of samples from the deeper core is also highly variable with average Mg/Ca values for individual image maps ranging from 2.8 – 3.7 mmol/mol with an average of 3.2 mmol/mol (N = 12) (~16% lower than the shallow core data). Again, the average Mg/Ca values are in good agreement with bulk Mg/Ca values (3.5 mmol/mol) obtained from HR-ICPMS analysis [24]. Some of the shells in samples from the deep core have chambers retain the banding present in the shallow core shells, but the intensity of the banding is diminished (Figure 1C); in other deep core shells banding is absent (Figure 1D). Histograms of the Mg/Ca ratios (on scans where chamber number could be identified) reveal a decrease in the frequency of measurements that exceed 6.5 mmol/mol and an increase in the frequency of lower Mg/Ca measurements in the deeper core samples in comparison to the shallow core samples (Figure 2).

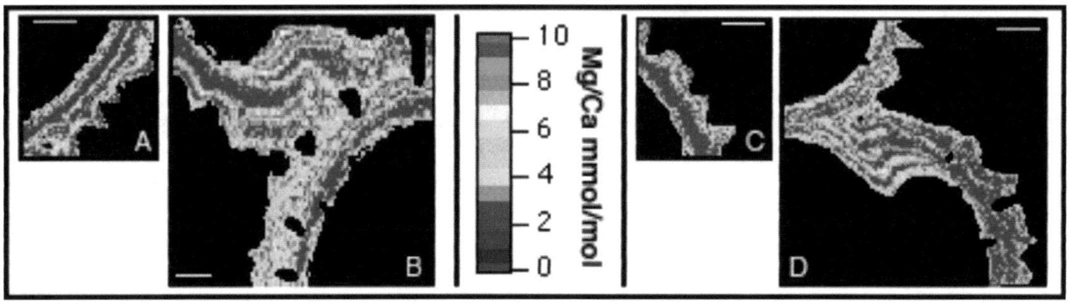

Figure 1: Representative images of *G. ruber* from the shallow (A and B) and deep (C and D) cores. Scale bar is 20μm.

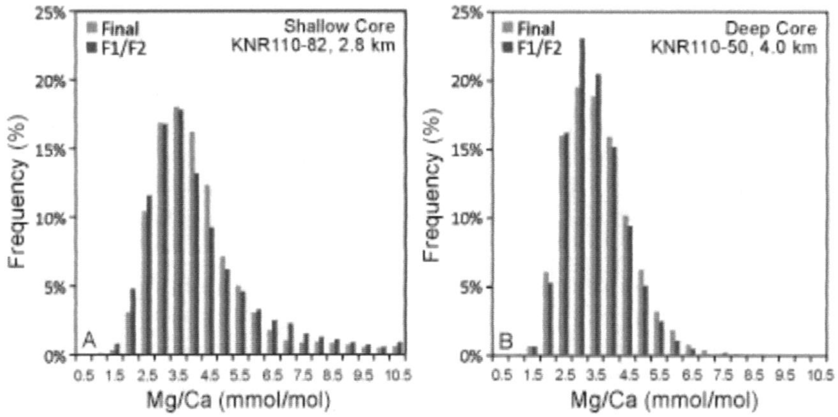

Figure 2: Histograms of *G. ruber* results from the shallow core (A) and deep core (B).

3.2. N. dutertrei results

EMPA image maps reveal *N. dutertrei* Mg/Ca distribution is highly variable within a single test and among shells obtained from the same interval. In some shells, there are thin alternating layers of high and low Mg/Ca (i.e. banding similar to that of *G. ruber*) and the average Mg/Ca ratio for a single image map is very high (>2.7 mmol/mol). In others shells, the Mg/Ca ratio is very low throughout the entire shell or is high in the inner portion of the chamber wall and low in the outer portion of the chamber wall.

In the shallow core, the average Mg/Ca values for individual image maps range from 1.5 to 3.2 with an average of 2.2 mmol/mol (N = 12) (Figures 4A and 4B). The average Mg/Ca ratio is in good agreement with bulk Mg/Ca values for these samples obtained from HR-ICPMS analysis (2.5 mmol/mol; [24]). In the deeper core, average the Mg/Ca ratio of individual maps range from 1.5 to 3.0 mmol/mol Mg/Ca with an average Mg/Ca ratio of 1.9 mmol/mol (N = 15) (~14% lower Mg/Ca in comparison to the shallow core) (Figures 4C and 4D). The average Mg/Ca ratio is in good agreement with published Mg/Ca values for these samples obtained from HR-ICPMS analysis (2.1 mmol/mol; [24]).

Identifying chamber number in cross-section is difficult in *N. dutertrei*, therefore rather than separating the samples by chamber number, histograms were generated on the low vs. high Mg/Ca samples. The histograms reveal the distribution of the Mg/Ca ratio in the high Mg/Ca samples is nearly identical in both the shallow and deep cores (Figure 4A). In the low Mg/Ca ratio samples the frequency of high Mg/Ca values decreases and low Mg/Ca ratio measurements increase in the deeper core (Figure 4B).

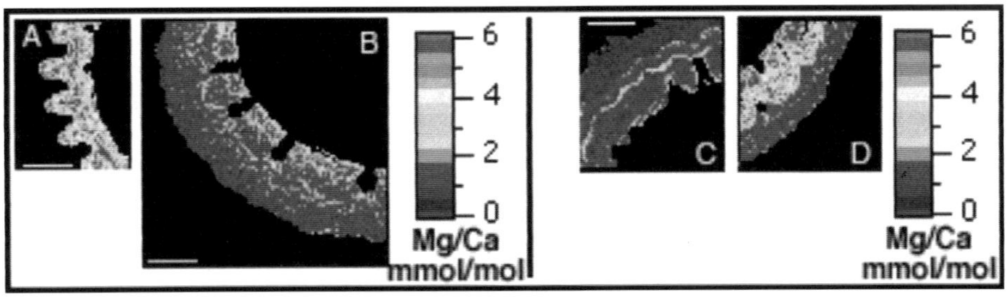

Figure 3: Representative image maps from the shallow (A and B) and deep (C and D) cores. Scale bar in maps is 20 microns. SEM images of scans A and B are detailed in Figure 3D and 3E.

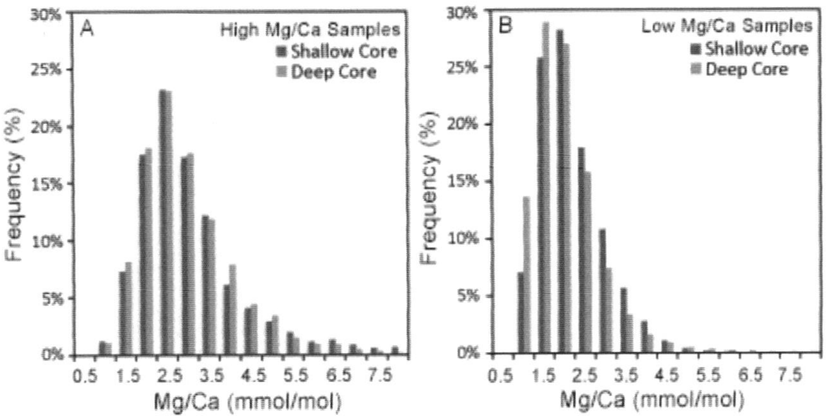

Figure 4: Histograms of low Mg/Ca shells (A) vs. high Mg/Ca shells (B) in both the shallow core (blue) and deep (red) cores.

4. Discussion and Conclusion

The shallow core images for *G. ruber*, a symbiont-bearing species, reveal distinct bands of high and low Mg/Ca layers consistent with previous studies [29]. The thin layers are thought to be calcite lain on the shell when a new chamber is formed [34, 35]. In deeper samples the banding appears to be diminished, though not uniformly, and in some specimens the banding is absent. Histograms generated on shallow and deep core samples reveal the frequency of measurements greater than 6.5 mmol/mol decreases in the deeper core samples and the frequency of lower mmol/mol values increases suggesting portions of the shell with high Mg/Ca calcite is preferentially lowered by dissolution though the mechanism by which the inner layer calcite is preferentially lowered by dissolution has not been determined.

The shallow core images for *N. dutertrei*, an asymbiotic species, show the Mg/Ca ratios of this species is highly variable, not only within a single chamber, but also among chambers in a single shell and among shells obtained from the same sample. For example, chamber-to-chamber average Mg/Ca ratios can differ by up to 35% in a single shell and the average Mg/Ca ratio of individual maps can vary by over 50% in samples obtained from the same sediment interval. Owing to the highly variable nature of the Mg/Ca ratio in this species, identifying the effects of dissolution has proved difficult. Histograms suggest, similar to *G. ruber*, that higher Mg/Ca ratios are lowered by dissolution.

SEM images may shed light on the differences in the Mg/Ca variability in *N. dutertrei*. SEM images reveal differences in surface texture and wall thickness (Figure 5A-5G). The thick walled shells have a highly crystalline thick 'crust' with well-defined euhedral calcite crystals on the surface and few exposed pores (Figures 5A and 5F). The thin walled shells lack euhedral surface crystals and have large exposed pores on all chambers (Figure 5B and 5G). This type of shell morphology for *Neogloboquadrinids* was reported in Sautter, 1998, who termed the difference in shell morphology as crystalline (highly crystalline) or reticulate (without the thick highly crystalline crust) [38]. Our SEMs reveal some shells have a transitional texture with both surface morphologies (Figure 3C). The wall thickness is highly variable; some shells have all thin walls and some have all thick walls (see for example Figures 5D and 5E). EMPA images obtained from the thin vs. thick shells suggest they have different Mg/Ca distributions (see image maps of *N. dutertrei* in Figures 3A and 3B which are highlighted in the white boxes in SEM Figures 5D and 5E). We hypothesize that the differences in the Mg/Ca ratio are due to different morphotypes of this species. Alternatively, the changes in shell surface and Mg/Ca ratio may reflect changes in calcification throughout the life cycle of *N. dutertrei*. Either hypothesis may explain the existence of a transitional texture from thin and thick shells through the life cycle of this species and a change in the Mg/Ca ratio preserved in the shell wall.

Results, though preliminary, suggest the EMPA technique can be used to characterize patterns of Mg/Ca variability in well-preserved samples that change with increased dissolution. The highly variable nature of Mg/Ca within individual tests makes it difficult to ascertain specific changes that occur with dissolution; however, some patterns are beginning to emerge. For example, the decrease in the high (>6.5 mmol/mol) Mg/Ca calcite in the deeper *G. ruber* samples, highlighted by the histograms of Mg/Ca values (Figure 2). Determining changes in the variability of the Mg/Ca ratio in *N. dutertrei* has proven difficult. Discerning the highly variable nature of the Mg/Ca ratio from chamber to chamber is complicated by the difficulty in identifying chamber number in cross section, which make it difficult to discern ontogenic changes in Mg/Ca distribution. Understanding the change in the Mg/Ca ratio due to dissolution will likely require additional maps of samples from other ocean basins, such as the deep Pacific where dissolution is intense. Additionally, obtaining images of shells from periods of time in the past when preservation and water mass chemistry were different from today will aid in determining more precise changes in shell Mg/Ca distribution that occur with dissolution.

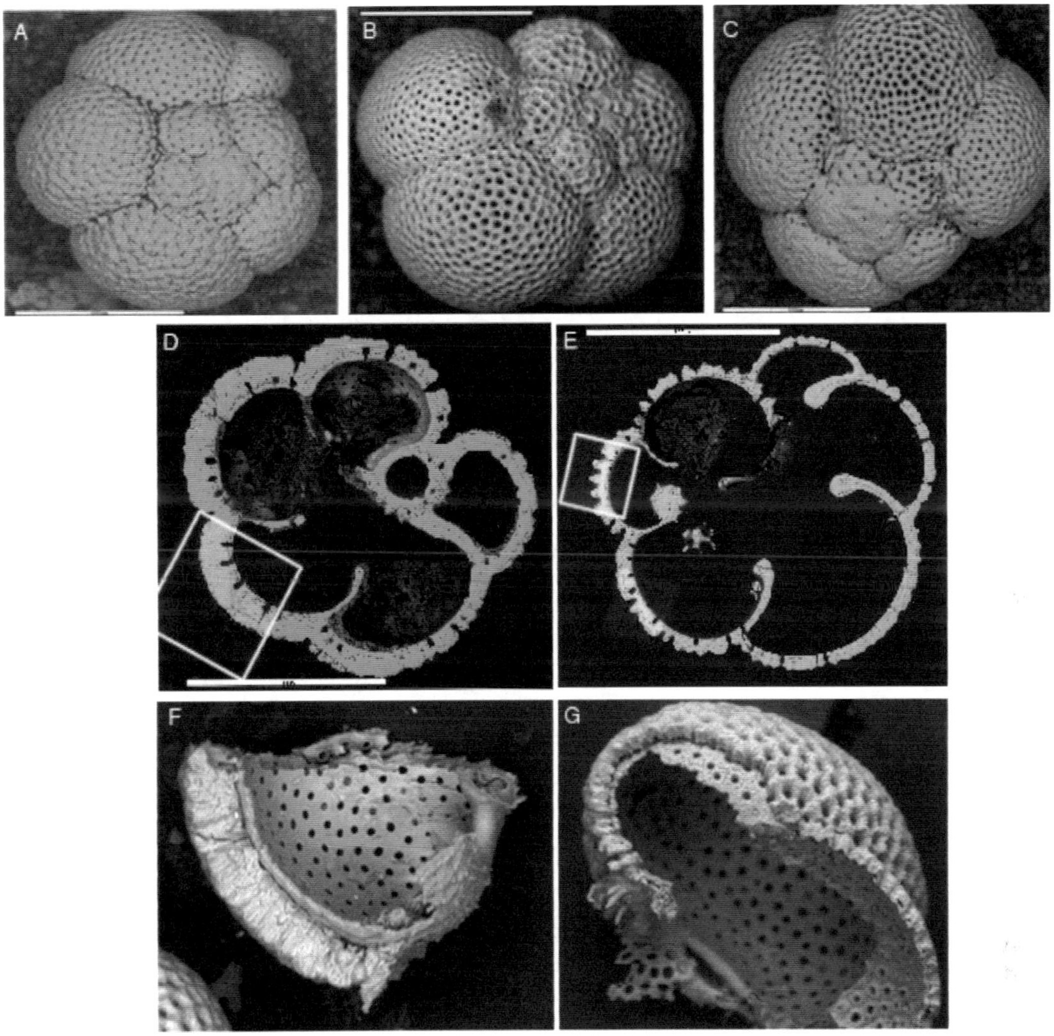

Figure 5: SEM images of *N. dutertrei*. (A) Highly crystalline surface morphology with a thick outer layer with few exposed pores. (B) Surface lacks crystalline outer crust and has large exposed pores on all chambers. (C) A transitional morphology that contains both the highly crystalline crust and chambers that lack the crystalline crust and retain exposed pores. (D) Cross-section of thin chambered *N. dutertrei*. (D) Cross section of thick-chambered *N. dutertrei*. (White boxes in (D) and (E) are the scans shown in Figure 2A and 2B above). (F) Fragment of shell with thick outer layer. (G) Fragment of shell with thin chamber walls and no thick outer layer.

Acknowledgments

We thank Ian Steele at the University of Chicago for assistance with the SEM and electron microprobe. Andy Davis from the University of Chicago for assistance with developing our mapping methodology and with data processing. Paul Craddock from the University of Chicago Origins Group assisted with analysis of the calcite spar. Assistance from Kim Knight from LLNL (previously of the Origins Group of the University of Chicago) improved sample preparation. Discussions with Frank Richter improved our analysis technique. Reviews from Aleksey Sadekov and Leopoldo Pena improved this manuscript.

References

[1] Oomori, T., K. Hiroshi, and Y. Maezato (1987), *Marine Chemistry*, *20*, 327-336.
[2] Katz, A. (1973), *Geochimica et Cosmochimica Acta*, *37*, 1563-1586.

[3] Lea, D. W., T. Mashiotta, and H. Spero (1999), *Geochim. et Cosmochim. Acta*, *63*(16), 2369-2379.
[4] Mashiotta, T., D. Lea, and H. J. Spero (1999), *Earth and Planetary Science Letters*, *170*, 417-432.
[5] Nurnberg, D., J. Bijma, and C. Hemleben (1996), *Geochim. et Cosmochim. Acta*, *60*(5), 803-814.
[6] von Langen, P. J., D. Pak, H. Spero, and D. Lea (2005), Geochem. Geophys. Geosyst., 6, Q10P03.
[7] Anand, P., H. Elderfield, and M. H. Conte (2003), *Paleoceanography*, *18*(2), 1050.
[8] McConnell, M. C., and R. C. Thunell (2005), *Paleoceanography*, *20*(9).
[9] McKenna, V. S., and W. L. Prell (2004), *Paleoceanography*, *19*(2), PA2006.
[10] Elderfield, H., and G. Ganssen, Nature, 405, 442–445, 2000.
[11] Cléroux, C., E. Cortijo, P. Anand, L. Labeyrie, F. Bassinot, N. Caillon, and J.-C. Duplessy (2008), *Paleoceanography*, 23, PA3214.
[12] Dekens, P. S., D. W. Lea, D. K. Pak, and H. J. Spero (2002), *Geochem. Geophys. Geosys.*, *3*(4).
[13] de Garidel-Thoron, T., Y. Rosenthal, L. Beaufort, E. Bard, C. Sonzogni, and A. C. Mix (2007), *Paleoceanography*, *22*, PA3204.
[14] Nürnberg, D., Müller, A., Schneider, R.R. (2006), *Paleoceanography*, 15(1), 124–134.
[15] Regenberg, M., D. Nürnberg, S. Steph, J. Groeneveld, D. Garbe-Schönberg, R. Tiedemann, and W.-C. Dullo (2006), *Geochem. Geophys. Geosyst.*, *7*.
[16] Lear, C., Elderfield, H., Wilson, P.A. (2000), *Science*, 287 (5451), 269–272.
[17] Hastings, D. W., A. Russell, and S. Emerson (1998), *Paleoceanography*, *13*(2), 161-169.
[18] Regenberg, M., S. Steph, D. Nurnberg, R. Tiedemann, and D. Garbe-Schonberg (2009), *Earth and Plan. Sci. Letters*, 279, 324-336.
[19] de Villiers, S. (2003), *Paleoceanography*, *18*(3), 1070-1079.
[20] Brown and Elderfield, *Paleoceanography,* 11 (1996), pp. 543–55.
[21] Peterson, M. N. A. (1966), *Science*, *154*, 1542-1543.
[22] Archer, D. (1996), Global Biogeochemical Cycles 10: 511-526.
[23] Rosenthal, Y., and G. P. Lohmann (2002), *Paleoceanography*, *17*(3)
[24] Fehrenbacher, J., P. Martin, and G. Eshel (2007), In: *AGU, Fall mtg 2007*, *Abstract #PP43#-07*.
[25] Fehrenbacher, J. S., M. P.A., and G. Eshel (2006), *Geochemistry Geophysics Geosystems*, *7*(9).
[26] Fehrenbacher, J.S., P.A. Martin, *in prep*
[27] Eggins, S.M., A. Sadekov, P. De Deckker (2004), *Earth and Plan. Sci. Letters*, 225, 411-419.
[28] Hathorne, E. C., O. Alard, R. James, and N. Rogers (2003), *Geochem. Geophys. Geosyst.*, 4(12).
[29] Sadekov, A. Yu., S. M. Eggins, and P. De Deckker (2005), *Geochem. Geophys. Geosyst.*, 6.
[30] Kunioka, D., K. Shirai, N. Takahata, Y. Sano, T. Toyofuku, and Y. Ujiie (2006), *Geochem. Geophys. Geosyst.*, 7.
[31] Levitus, 1994; http://iridl.ldeo.columbia.edu/sources/.levitus94/.annual/.temp/
[32] Klein, C., American Journal of Science (1960), 258, 402.
[33] Pena, L. D., I. Cacho, E. Calvo, C. Pelejero, S. Eggins, and A. Sadekov (2008), *Geochem. Geophys. Geosyst.*, 9.
[34] Bé, A., C. Hemleben, O. Anderson, and M. Spindler (1979), *Micropaleontology*, *25*, 294–307.
[35] Hathorne, E. C., R. H. James, and R. S. Lampitt (2009), *Paleoceanography*, *24*, PA4204,

North Atlantic *Globorotalia inflata* coretop Mg/Ca calibrations and temperature reconstructions over Termination I

Elizabeth J Farmer, Mark R Chapman and Julian E Andrews

University of East Anglia, School of Environmental Sciences, Norwich, NR4 7TJ

e.farmer@uea.ac.uk

Abstract. Mg/Ca ratios from planktonic foraminifera are widely used as an independent proxy for ocean temperature. The choice of calibration is vital for reconstructing meaningful temperatures. We present preliminary results based on a new Mg/Ca-temperature calibration ($Mg/Ca = 0.92e^{0.039T}$) for the thermocline-dwelling foraminifera *Globorotalia inflata*, derived using a suite of North Atlantic coretops. The downcore Mg/Ca record from core SU90-03 (40°N, 32°W) was used to compare how our new and existing calibrations performed for the Holocene and over the large scale change associated with Termination I.

1. Introduction

The measurement of Mg/Ca ratios in the tests of foraminifera is now a routine technique for estimating past ocean temperatures [1]. Advantages over other palaeothermometry techniques are the capacity to measure Mg/Ca on the same sample as $\delta^{18}O$, allowing the potential for palaeosalinity reconstructions, and the avoidance of non-modern-analogue issues that can be associated with the faunal assemblage based transfer function approach. Use of the Mg/Ca proxy has developed over the last decade and multi-species reconstructions based on species-specific temperature calibrations are now increasingly being employed [2-4]. This enables a profile of the water column properties at a particular site to be built up by comparing species from different depth habitats over time. This approach has huge potential for palaeoceanographic studies aiming to reconstruct past changes of different water masses.

More recently, research has begun to highlight complicating factors in the use of Mg/Ca and the key assumption that it records solely a temperature signal. Such factors include: sensitivity to salinity and seawater carbonate ion concentration [5-7]; partial dissolution, either where deep cores approach the carbonate compensation depth, with different species being affected to a greater or lesser degree [8, 9] or due to differences in the aggressiveness of the various cleaning routines applied to the samples [10]. The ecology of the species under study must also be taken into consideration when interpreting Mg/Ca records, since the limits of a species' environmental niche can vary – aspects such as seasonality and depth habitat may alter across a species' biogeographic range and through time. Whilst these factors need to be considered, discounted or corrected for, the reliability of the Mg/Ca temperature reconstruction is dependent ultimately on the calibration used to convert raw Mg/Ca ratios to temperatures. Due to the significant differences in Mg incorporation between species and locations [2, 3], detailed multi-species studies require accurate, species-specific and even region-specific calibrations.

Here we use a suite of coretop data from the North Atlantic to derive a new Mg/Ca-temperature calibration for the thermocline-dwelling species *Globorotalia inflata*. A downcore Mg/Ca record is then used as a case study to explore the existing temperature calibrations for this species and highlight the importance of calibration selection in the interpretation of Mg/Ca records.

2. Core sites and methods

This study was conducted using box core surface samples collected during the North East Atlantic Palaeoceanography and Climate Change (NEAPACC) program, along with coretop and downcore sediment samples from core SU90-03 (40° 30'N, 32° 03'W; 2475 m water depth), a piston core collected from the northern reaches of the subtropical gyre in the North Atlantic (Figure 1). Core SU90-03 is from a key area for monitoring the strength of the North Atlantic Current (NAC) during both glacial and interglacial periods, as previous work from this core indicated that this site remained south of the polar front even during the last glacial maximum (LGM) [11, 12]. The location of the core also allows a relatively rare opportunity to obtain a continuous Mg/Ca record from a single species across the major climatic transition of Termination I since *G. inflata* relative abundances show that this species is present throughout the last glacial-interglacial climate [12].

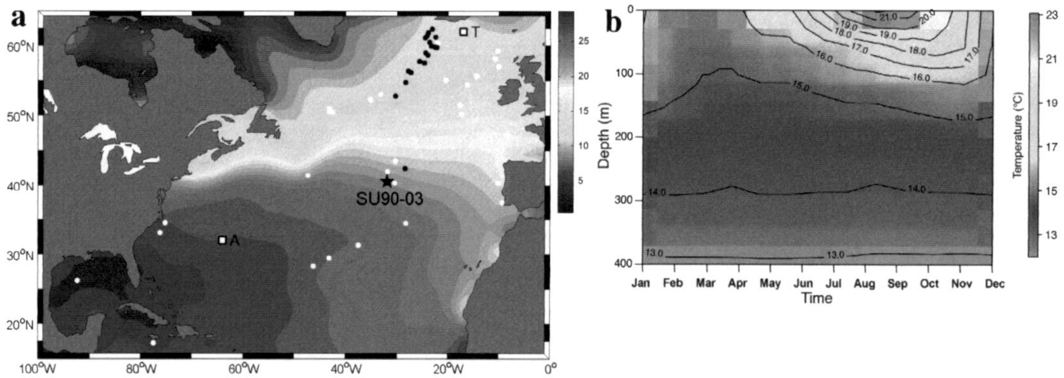

Figure 1: a. Core locations: NEAPACC cores in black, Cléroux *et al.* [3] coretops in white. Contours show August SST in the North Atlantic [13]. Core SU90-03 (used in this study and that of Cléroux *et al.* [3]) is marked, along with the sites used in the Anand *et al.* [2] (A) and Thornalley *et al.* [4] (T) studies; b. SU90-03 core site hydrography based on WOA98 [13].

At present, *G. inflata* is most abundant in temperate waters associated with the North Atlantic Current (NAC) but also ranges to subpolar and subtropical waters, with a temperature range of 8-18 °C. It is a deep dwelling planktonic species associated with the base of the seasonal thermocline [14, 15], which at the mid-latitudes sites considered in this study is located at approximately 100m depth. The modern day hydrography at the site of core SU90-03 (Figure 1b) illustrates the development and deepening of the seasonal thermocline during summer months. Temperatures between the surface and 100 m range from 23 °C to 15 °C, with summer thermocline temperatures ranging from ~15-16 °C. There are a number of Mg/Ca-temperature calibrations for *G. inflata* but the available data is limited, particularly at the cooler end of the species temperature range. In order to address this and create a calibration suitable for widespread usage across the mid-latitude North Atlantic, we analyzed *G. inflata* (300-355μm) in 18 NEAPACC coretops for both $\delta^{18}O$ and Mg/Ca.

3. Methods and Instrumentation

Mg/Ca analyses by ICP-AES were conducted using the intensity ratio calibration method of de Villiers *et al.* [16] and using the cleaning routine of Barker *et al.* [10]. Calibration standards were prepared from high purity $CaCO_3$ (NIST 915b), $SrCO_3$ (NIST 987) and Mg metal rod. Levels of contaminants for each were in agreement with the assay certificates. Analysis of the limestone reference standard ECRM 752-1, mean Mg/Ca = 3.97 mmol/mol, are consistent with results published by Greaves *et al.* [17] and interlaboratory reproducibility identified by Rosenthal *et al.* [18]. The ECRM 752-1 and two other solutions were used as internal consistency standards and were analyzed several times throughout each run to establish analytical reproducibility over time. The mean standard deviation of the Mg/Ca ratios (mmol/mol) for the internal standard is 0.027, with a mean relative standard deviation (RSD) of 0.49. When calculated with respect to temperature (°C) the mean standard deviation is 0.049, with a mean RSD of 0.25. Further information on instrumentation are detailed in Farmer *et al.* [19]. Uncertainties associated with the calibration to temperature (see Section 4) are far greater than the intra-sample differences encountered for replicate foraminiferal Mg/Ca measurements.

Parallel measurements of the $\delta^{18}O$ of foraminiferal calcite ($\delta^{18}O_f$) are used to estimate calcification temperature and calibrate the Mg/Ca palaeothermometer. Isotopic temperatures in this study were calculated for the coretops using the palaeotemperature equation of Shackleton [20]:

$$Tiso = 16.9 - 4.38(\delta^{18}O_f - \delta^{18}O_{sw}) + 0.1(\delta^{18}O_f - \delta^{18}O_{sw})^2 \quad (1)$$

where $\delta^{18}O_f$ is the $\delta^{18}O$ from the foraminiferal calcite and $\delta^{18}O_{sw}$ is the $\delta^{18}O$ of the seawater. This is consistent with the methods used by Cléroux *et al.* [3] and Anand *et al.* [2], allowing for direct comparison of our results. Errors associated with isotopic temperatures (Tiso) are estimated to be of the order of ± 1.5 °C and take into account uncertainties primarily related to the estimation of $\delta^{18}O_{sw}$ [3, 15] and the selection of temperature equation used in the calculation of Tiso.

We then used these coretop data to compare with three published calibrations for *G. inflata* [2-4]. The coretop SU90-03 is included both in our new coretop dataset and in the Cléroux *et al.* [3] study. The similarity of the independently analyzed results (Figure 2) provides further confidence in the consistency and reliability of the new coretop dataset.

4. Coretop calibration

Existing calibrations specific to *G. inflata* differ significantly in their methodology. The Anand *et al.* [2] calibration is based on sediment trap samples from the Sargasso Sea (32°05.4'N, 64°15.4'W) and uses a slightly modified cleaning routine to overcome the higher levels of organic matter often found in sediment trap samples in comparison to typical sediment core material. *G. inflata* specimens from the 350-500 µm size fraction produced Mg/Ca ratios of 1.5-2.0 mmol/mol and a Tiso range of 17–21 °C. The *G. inflata* calibration of Cléroux *et al.* [3] is based on 29 coretops from across the North Atlantic ranging from 60 °N to 17 °N, including the SU90-03 coretop, covering a range of Mg/Ca ratios of 1.1-2.0 mmol/mol and a Tiso range of 10.5-17.5 °C. Specimens from the 250-315 µm and 355–400 µm size fractions were analysed but since there was no evidence of a significant size effect on trace metal ratios, the two datasets were combined. However, test size is known to affect $\delta^{18}O$ values and thereby Tiso, and Cléroux *et al.* [3] suggest that this could explain the differences between their own calibration and that of Anand *et al.* [2]. Different methodologies and the different types of sample material (sediment trap versus coretop sediments) are also likely play a part. The Thornalley *et al.* [4] calibration was generated to study the downcore record of the subpolar North Atlantic core RAPiD 12-1K and is derived from the coretop Mg/Ca and $\delta^{18}O$ of the same core, most likely making it a more subpolar- and location-specific calibration. No raw Mg/Ca data is presented in Thornalley *et al.* [4] but the reconstructed Holocene temperatures at the core site equate to Mg/Ca ratios of approximately 1.6-2.0 mmol/mol. The methodology in the present study is most directly comparable to the Cléroux *et al.* [3] study, although specimens from a different size fraction were analysed (300-355 µm), which could explain the offset in the $\delta^{18}O$ between the two SU90-03 coretop points as highlighted in Figure 2.

Results from our coretop analyses are plotted alongside the three published Mg/Ca-temperature calibration curves in Figure 2, with the coretops of Cléroux *et al.* [3] and the additional NEAPACC and SU90-03 coretops shown. We combined our new high latitude suite of coretops with the Cléroux *et al.* [3] coretops to generate a new calibration curve. In order to maintain consistency with our new dataset, where appropriate we calculated the mean for individual cores in the Cléroux *et al.* [3] dataset such that each location has equal weighting in the regression. The new calibration, based on the combined dataset, is:

$$Mg/Ca = 0.916 \pm 0.07 e^{0.039 \pm 0.006 T} \quad (r^2 = 0.52) \quad (2)$$

IOP Conf. Series: Earth and Environmental Science **9** (2010) 012019 doi:10.1088/1755-1315/9/1/012019

This calibration extends the lower limit of the *G. inflata* temperature range to 8 ºC, although the best fit calibration curve does not differ markedly from the relationship defined in the Cléroux *et al.* [3] calibration. In comparison, the curve describing the Anand *et al.* [2] calibration in Figure 2, shows a shift to lower Mg/Ca or higher Tiso – for the reasons previously discussed, suggesting higher Mg/Ca derived temperatures. The Thornalley *et al.* [4] calibration curve departs markedly from the low exponential of the other three calibrations (due to the higher exponent in their calibration: 0.1 rather than 0.058, 0.056 and 0.039 for Anand *et al.* [2], Cléroux *et al.* [3] and this study, respectively). The Thornalley *et al.* [4] calibration involves much higher Mg/Ca ratios for any given Tiso, such that to generate a temperature of 15 ºC, a Mg/Ca ratio in excess of 3.0 mmol/mol would be necessary, rather than the 1.8 mmol/mol required by both the Cléroux *et al.* [3] calibration and the new calibration presented here, or the 1.25 mmol/mol of the Anand *et al.* [2] calibration. This calibration is thus likely to return markedly cooler temperatures than the other calibrations for the range of Mg/Ca ratios seen in our coretops.

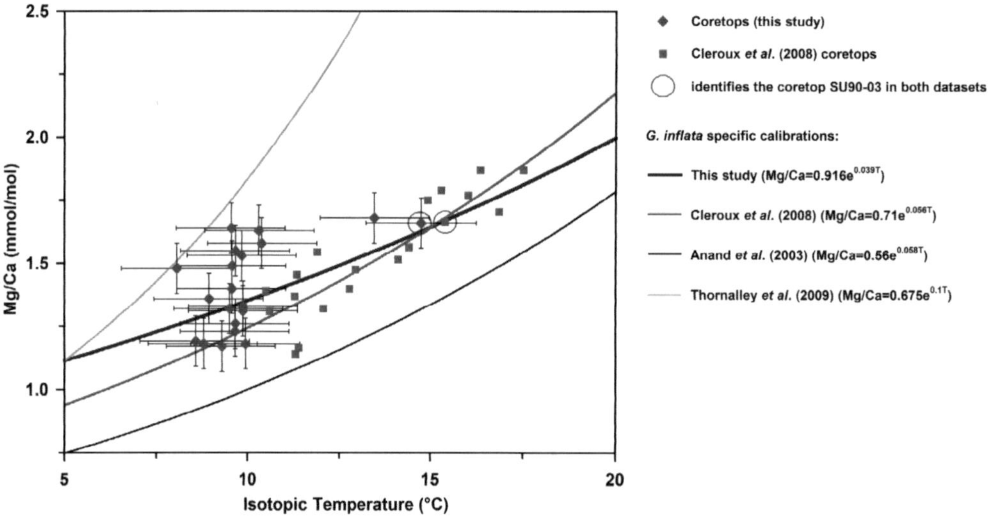

Figure 2: *G. inflata* Mg/Ca–temperature calibrations. Purple points mark the new coretop data from this study, green points mark the coretops used in the Cléroux *et al.* [3] study. The uppermost pale grey line shows the calibration curve of Thornalley *et al.* [4]; the lower black line shows the Anand *et al.* [2] calibration. The dark grey line is the Cléroux *et al.* [3] calibration curve, overlapping with the new calibration curve in thick blue. Red circles identify coretop data for SU90-03 from both datasets.

The new coretops show a relatively large scatter at lower temperatures which reduces the r^2 of the calibration to 0.52, from the 0.72 of the original Cléroux *et al.* [3] calibration. Further work is needed to resolve this scatter, in the form of additional Mg/Ca and $\delta^{18}O$ replicates and further assessment of the quality of the coretop material. Since these temperatures are at the very extreme of *G. inflata* temperature tolerance, it may well be that this scatter represents populations under stress, such that ecological factors including discontinuous productivity and changing seasonality and depth habitat become increasingly significant. For the new calibration, the uncertainties associated with the temperature estimates were calculated by looking at the difference between the Mg/Ca derived temperature and Tiso values. The mean of all the differences is 2 ºC and the standard deviation on the differences is 1.5 ºC. Errors quoted by Cléroux *et al.* [3] and Thornalley *et al.* [4] are 1.4 ºC and 1.3 ºC respectively.

5. Application of Mg/Ca-temperature calibrations

The results obtained when applying the alternative *G. inflata* Mg/Ca-temperature calibrations to the downcore Mg/Ca record from core SU90-03 are shown in Figure 3. The shaded bar represents the modern day temperature at the core site at 100 m water depth, i.e. near the supposed *G. inflata* calcification depth at the base of the summer thermocline. Error bars associated with each calibration are drawn on the right to give an indication of the statistical significance of the differences between the reconstructed temperatures. The various calibrations produce values ranging from 5 °C to 11 °C for the maximum magnitude of temperature shift between full glacial and interglacial conditions. These results demonstrate how the choice of calibration plays a key role in determining not only individual temperature values, but also the amplitude of the relative change in temperature over time. Such a divergence in results is not helpful and needs to be better constrained if meaningful palaeoceanographic interpretations are to be drawn.

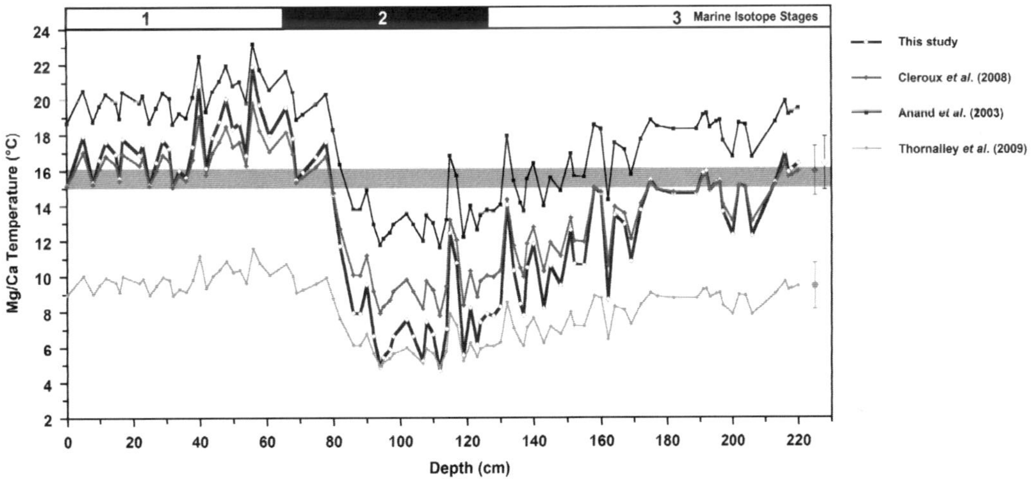

Figure 3. Mg/Ca temperature reconstructions from Anand *et al.* [2] (topmost black line), Thornalley *et al.* [4] (lower pale grey line) and Cléroux *et al.* [3] (mid grey line). The reconstruction based on the new calibration is shown in blue. Shaded bar indicates modern summer thermocline temperatures at the core site (100 m depth).

The Mg/Ca based temperature reconstructions vary significantly between the calibrations, although, as expected, the new calibration and the original Cléroux *et al.* [3] calibration are largely within error of each other. The only interval where these two deviate notably is during the coldest periods, during the last glacial maximum (LGM), with the new calibration producing values as much as 3 °C lower than the Cléroux *et al.* [3] calibration. The robustness of both these calibrations for modern day temperature ranges is indicated by their ability to correctly reconstruct modern day thermocline temperatures at the core site (shaded bar). The Anand *et al.* [2] calibration produces the highest temperatures, with Holocene estimates notably higher then modern day thermocline temperatures and actually closer to modern day surface temperatures. In places the derived temperatures exceed the species' maximum temperature range reflected in the coretop Tiso measurements (Figure 2). The Thornalley *et al.* [4] calibration produces the coldest reconstruction, with relatively muted variability and with a maximum temperature almost 11 °C cooler than that of Anand *et al.* [2]. Interestingly, the new calibration draws closer to the Thornalley *et al.* [4] calibration at the cold extreme. This suggests that the new calibration is capable of reconstructing both the subtropical conditions of the present day and the much cooler conditions of the last glacial period for this region. However, the lowest temperatures of 5 °C are below the lower limit of the species temperature range in the coretops (taking into account the calibration errors). More work is needed to constrain the extent of the temperature changes and assess whether other factors are asserting a significant influence on the glacial-interglacial shift in Mg/Ca ratios.

In spite of the differences in the temperature records, it is clear that thermocline temperature changes at the site of core SU90-03 accompanied the build-up of the ice sheets during the last glacial period and the subsequent disintegration of the ice sheets following the LGM. Early marine isotope stage (MIS) 3 exhibits maximum temperatures that are not too dissimilar to modern conditions. Thermocline temperatures appear to steadily decline towards the LGM, punctuated by a number of brief warmings, reaching a minima of ~ 5 °C, according to the new calibration. The warming into the Holocene is rapid and step-like and suggests a period of early Holocene warmth at depth before temperatures at the thermocline cooled and the modern day conditions of ~16 °C were established. This record provides a rare opportunity to study a single species' response to large scale climatic change in the northern subtropical gyre. Further work and additional proxies are required to develop a better understanding of glacial to interglacial changes in this region and to examine the relationship between thermocline and near surface temperature histories during large scale events and under different climatic regimes.

6. Conclusions

We define a new Mg/Ca-temperature calibration ($Mg/Ca = 0.92e^{0.039T}$) for the thermocline-dwelling foraminifer *G. inflata* based on a series of North Atlantic coretops. This calibration builds upon and is largely consistent with the work of Cléroux *et al.* [3] although it appears to better estimate the cooler temperatures associated with the lower limit of *G. inflata's* temperature range. The suitability of the new coretop calibration is supported by its ability to reconstruct the present day temperatures at the site of core SU90-03. Two other calibrations were compared in order to highlight how the choice of calibration can have a significant impact on the interpretation of the Mg/Ca record. A comparison of the reconstructed temperatures using downcore data from North Atlantic core SU90-03 demonstrates that the choice of calibration affects not only the absolute temperatures but the amplitude of events, particularly over large scale climatic transitions such as Termination I. This potentially has major implications for errors associated with reconstruction of the $\delta^{18}O$ of the seawater/palaeosalinity and surface temperature gradients based on Mg/Ca records from multiple foraminiferal species.

Acknowledgements

Technical support on the ICP-AES was provided by G. Chilvers. Thanks to D. Sproson for help with mapping and Matlab plotting. This research was funded through the NERC RAPID program (grant NE/C50907/1). Attendance of the PAGES 1st YSM and 3rd OSM in July 2009 was supported by QUEST. The manuscript benefitted from comments from two anonymous reviewers and YSM guest editor C. Cléroux.

References

[1] Elderfield H and Ganssen G 2000 Past temperature and δ18O of surface ocean waters inferred from foraminiferal Mg/Ca ratios Nature 405 442-5.
[2] Anand P, Elderfield H and Conte M H 2003 Calibration of Mg/Ca thermometry in planktonic foraminifera from a sediment trap time series *Paleoceanography* **18** 28-1. doi: 10.1029/2002PA000846
[3] Cléroux C, Cortijo E, Anand P, Labeyrie L, Bassinot F, Caillon N and Duplessy J C 2008 Mg/Ca and Sr/Ca ratios in planktonic foraminifera: Proxies for upper water column temperature reconstruction *Paleoceanography* **23** doi: 10.1029/2007pa001505
[4] Thornalley D J R, Elderfield H and McCave I N 2009 Holocene oscillations in temperature and salinity of the surface subpolar North Atlantic *Nature* **457** 711-4. doi: 10.1038/nature07717
[5] Nürnberg D, Bijma J and Hemleben C 1996 Assessing the reliability of magnesium in foraminiferal calcite as a proxy for water mass temperatures *Geochimica et Cosmochimica Acta* **60** 803-14.
[6] Lea D W, Mashiotta T A and Spero H J 1999 Controls on magnesium and strontium uptake in planktonic foraminifera determined by live culturing *Geochimica et Cosmochimica Acta* **63** 2369-79.

[7] Ferguson J E, Henderson G M, Kucera M and Rickaby R E M 2008 Systematic change of foraminiferal Mg/Ca ratios across a strong salinity gradient *Earth and Planetary Science Letters* **265** 153-66. doi: 10.1016/j.epsl.2007.10.011

[8] Dekens P S, Lea D W, Pak D K and Spero H J 2002 Core top calibration of Mg/Ca in tropical foraminifera: Refining paleotemperature estimation *Geochemistry Geophysics Geosystems* **3** doi: 10.1029/2001GC000200

[9] Mekik F, Francois R and Soon M 2007 A novel approach to dissolution correction of Mg/Ca-based paleothermometry in the tropical Pacific *Paleoceanography* **22** doi: 10.1029/2007PA001504

[10] Barker S, Greaves M and Elderfield H 2003 A study of cleaning procedures used for foraminiferal Mg/Ca paleothermometry *Geochemistry Geophysics Geosystems* **4** doi: 10.1029/2003GC000559

[11] Chapman M R and Shackleton N J 1998 Millennial-scale fluctuations in North Atlantic heat flux during the last 150,000 years *Earth and Planetary Science Letters* **159** 57-70.

[12] Chapman M R, Shackleton N J and Duplessy J C 2000 Sea surface temperature variability during the last glacial- interglacial cycle: assessing the magnitude and pattern of climate change in the North Atlantic *Palaeogeography Palaeoclimatology Palaeoecology* **157** 1-25.

[13] NODC WOA98. http://www.esrl.noaa.gov/psd/

[14] Cléroux C, Cortijo E, Duplessy J C and Zahn R 2007 Deep-dwelling foraminifera as thermocline temperature recorders *Geochemistry Geophysics Geosystems* **8** doi: 10.1029/2006gc001474

[15] Ganssen G M and Kroon D 2000 The isotopic signature of planktonic foraminifera from NE Atlantic surface sediments: implications for the reconstruction of past oceanic conditions. *J. Geol. Soc. Lond.* **157** 693–9.

[16] de Villiers S, Greaves M and Elderfield H 2002 An intensity ratio calibration method for the accurate determination of Mg/Ca and Sr/Ca of marine carbonates by ICP-AES *Geochemistry Geophysics Geosystems* **3** doi: 10.1029/2001GC000169

[17] Greaves M, Barker S, Daunt C and Elderfield H 2005 Accuracy, standardization, and interlaboratory calibration standards for foraminiferal Mg/Ca thermometry, *Geochemistry, Geophysics, Geosystems* **6** doi: 10.1029/2004GC000790.

[18] Rosenthal Y et al., 2004 Interlaboratory comparison studyof Mg/Ca and Sr/Ca measurements in planktonic foraminifera for paleoceanographic research *Geochemistry, Geophysics, Geosystems* **5** doi: 10.1029/2003GC000650.

[19] Farmer E J, Chapman M R and Andrews J E 2008 Centennial-scale Holocene North Atlantic surface temperatures from Mg/Ca ratios in Globigerina bulloides *Geochemistry Geophysics Geosystems* **9** doi: 10.1029/2008gc002199

[20] Shackleton N J 1974 Attainment of isotopic equilibrium between ocean water and the benthonic foraminifera genus *Uvigerina*: isotopic changes in the ocean during the last glacial *Cent. Natl. Rech. Sci. Colloq. Int.* **219** 203-10.

IOP Conf. Series: Earth and Environmental Science **9** (2010) 012020 doi:10.1088/1755-1315/9/1/012020

What caused *G. truncatulinoides* to calcify in shallower water during the early Holocene in the western Atlantic / Gulf of Mexico?

Caroline Cléroux[1,2] and Jean Lynch-Stieglitz[1]

[1] School of Earth and Environmental Sciences, Georgia Institute of Technology, Atlanta, GA, USA

[2] now at Lamont-Doherty Earth Observatory of Columbia University, Palisades, NY, USA

ccleroux@ldeo.columbia.edu

Abstract. In a recent paper we reconstructed the calcification depth of *G. truncatulinoides* since the last deglaciation in the Florida Straits. We highlighted a significantly different calcification depth between 8-10 ka and the late Holocene. This migration was found in three different cores over the western North Atlantic American continental margin. This paper presents new Mg/Ca ratio data on *G. truncatulinoides* in the Florida Strait and new stable isotopic measurements on *G. crassaformis* in the Gulf of Mexico. Preliminary interpretation to understand the cause of *G. truncatulinoides* migration in the water column at the beginning of the Holocene is given.

1. Introduction

Globorotalia truncatulinoides is a deep-dwelling planktonic foraminifera known to calcify around 200 meters depth in the western North Atlantic [5, 7, 8]. Paleoceanographic studies have used this species to reconstruct past thermocline condition assuming a constant calcification depth [14].

By comparing oxygen isotopic composition ($\delta^{18}O$) of surface dwelling and benthic foraminifera from cores along a depth transect [13], with *G. truncatulinoides* $\delta^{18}O$ in one of the deeper cores (KNR166-2-29JPC), we determined the calcification depth of this species over the past 12 ka in the Florida Straits (figure 1) [6]. This study showed that between 8 and 10 ka, *G. truncatulinoides* was calcifying in a much shallower habitat, around 130 m depth, than during the deglaciation or the late Holocene. We also showed that the large $\delta^{18}O$ signal associated with this migration is found in two other cores: RC12-10 in the western Gulf of Mexico and MD99-2203 off Cape Hatteras (figure 2). This is therefore a regional phenomenon over the western Atlantic continental slope that must have been triggered by environmental changes. The cause for this change of habitat is still unclear.

All the cores discussed in this paper are well dated. Cores KNR166-2-29JPC, RC12-10 and MD99-2203 have respectively 8, 7 and 11 radiocarbon dates over the period covered by this study [6, 15].

Here we present new data both on the Florida Straits core and on the Gulf of Mexico core to understand the context of this migration. Beyond elucidate the particular early Holocene migration in the western Atlantic margin region, we aim to constrain what parameters influence *G. truncatulinoides* calcification depth.

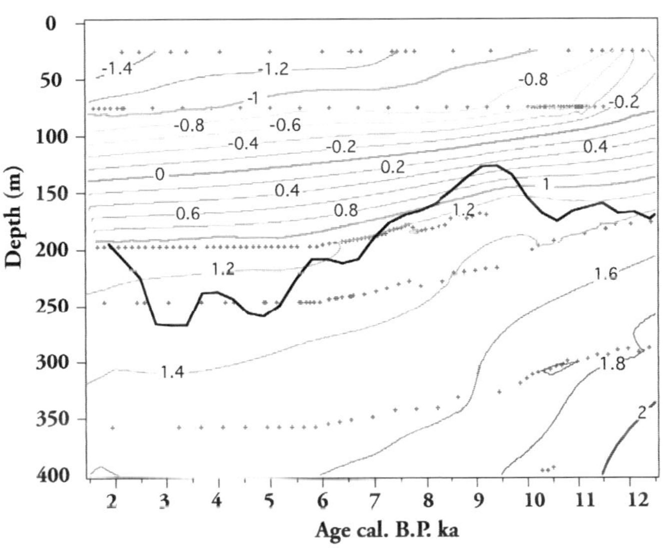

Figure 1: *G. truncatulinoides* calcification depth change in the Florida Straits over the last 12.5 ka (in calendar age B.P.). Red points: Pointer for $\delta^{18}O$ values at known water depth, from top to bottom: *G. ruber* (25m), *G. sacculifer* (75m), benthic foraminifera (modern depth : 198, 247, 358 and 446m). Thin colored lines: $\delta^{18}O$ isolines calculated by linear interpolation between $\delta^{18}O$ pointer values (red points). Thick black line: $\delta^{18}O$ and calcification depth of *G. truncatulinoides* from KNR166-2 29JPC measurements, $\delta^{18}O$ curve smoothed on 3 points.

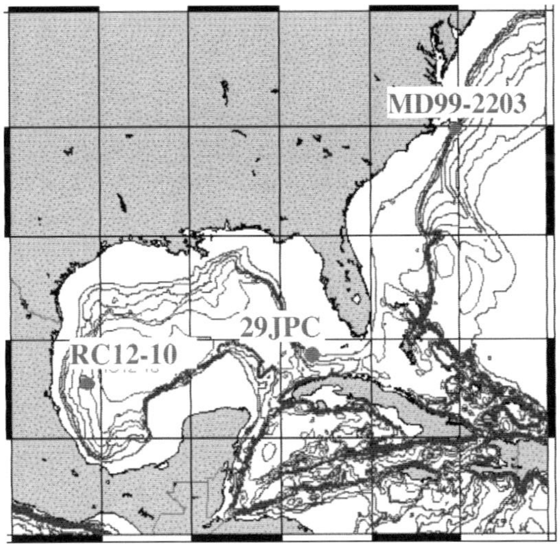

Figure 2: Location of the cores discussed in this paper.

2. New data and results

2.1. Mg/Ca ratio measurement of *G. truncatulinoides* in the core KNR166-2-29JPC and interpretation

About 10 tests of *G. truncatulinoides* dextral in the 355-500 μm fraction has been picked in the core KNR166-2-29JPC (24°16 N, 83°16 W, 648 m) for Mg/Ca ratio measurement. We used the inductively coupled plasma atomic-emission spectrometers (ICP-AES), model Ultima-C Jobin Yvon of Georgia

Institute of Technology. Inter-laboratories comparison of Mg/Ca measurements has been checked by standard powder analyses [9]. Analyses of standard powder are typically within the 2 % error for a mean Mg/Ca ratio of 5.55 mmol/mol.

The Mg/Ca ratio of foraminifera test is now widely accepted as a paleotemperature proxy. Specimens for $\delta^{18}O$ analysis and Mg/Ca ratio measurement were picked in the same samples. $\delta^{18}O$ value and temperature from Mg/Ca ratio are substituted into the oxygen isotope paleotemperature equation [16], to estimate seawater $\delta^{18}O$ using the following equations:

$$T_{Mg/Ca} = \frac{1}{0.074} \times \ln\left(\frac{Mg/Ca}{0.62}\right) \quad \text{equation 1}$$

$$\delta^{18}O_{seawater} = \delta^{18}O_{foram} - 5\sqrt{4.38 - \left(4.38 - 0.4 \times \left(16.9 - T_{Mg/Ca}\right)\right)} \quad \text{equation 2}$$

Equation 1 is the temperature – Mg/Ca ratio relationship for *G. truncatulinoides* [4]; equation 2 is the modified paleotemperature equation from [16], we applied a 0.27 correction to convert $\delta^{18}O_{foram}$ on the SMOW scale.

Temperature reconstruction and seawater $\delta^{18}O$, corrected for past ice volume change using the reconstruction of [18], are shown on figure 3. Estimated error on temperature reconstruction from *G. truncatulinoides* is 1.4°C; based on a simple error propagation, error on $\delta^{18}O_{seawater}$ reconstruction is about 1‰. Seawater $\delta^{18}O$ is regionally correlated to salinity, however we did not perform quantitative salinity calculation as the relationship may not be valid back in time.

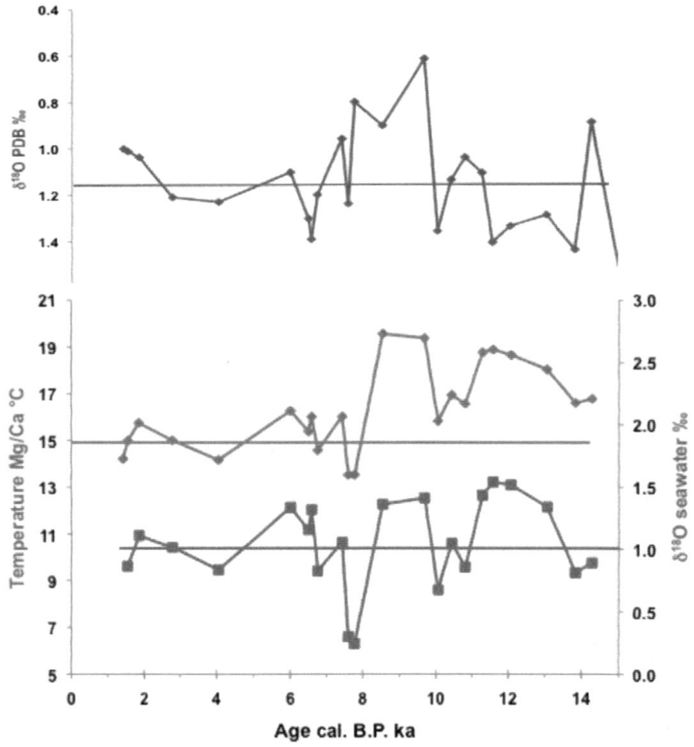

Figure 3: $\delta^{18}O$ PDB ‰ (green), temperature from Mg/Ca ratio (blue) and seawater $\delta^{18}O$ (red, corrected for sea level change) from *G. truncatulinoides* in the Florida Strait core KNR166-2-29JPC. The large $\delta^{18}O$ signal is also found in cores RC12-10 and MD99-2203 where *G. truncatulinoides* were picked in narrower size fractions.

Temperature reconstructions of *G. truncatulinoides* calcification habitat show a large step-like decrease of about 4°C at 8 ka (figure 3). The $\delta^{18}O_{seawater}$ doesn't show this step-like transition. The mean early Holocene $\delta^{18}O_{seawater}$ is about 0.3‰ higher than the value for the late Holocene (i.e. about 0.6‰ salinity, taking the mean $\delta^{18}O_{seawater}$/salinity relationship for the Western North Atlantic region between 0 and 500 meters depth). Contemporaneous to the temperature drop, the $\delta^{18}O_{seawater}$ show a large event of about 0.7‰ at 8 ka. Although the chronology differs somewhat, this depletion corresponds probably to the surface water event documented by [11] around 8.4 ka. These authors estimated that it reflects a salinity drop of about 5.5‰ and hypothesize that it precedes the widespread 8.2 ka event. Our data make a discussion about the chronology of these events difficult but it shows that the freshwater discharge reached thermocline depth. Clearly more investigations are needed to understand the cause and impact of such event.

2.2. $\delta^{18}O$ composition of *G. crassaformis* in the core RC12-10 and interpretation

Globorotalia crassaformis is also a deep-dwelling planktonic foraminifera species. Modern observation in Cariaco Basin indicates an habitat similar to *G. truncatulinoides* between 100-500 m [17]. Optimum conditions for *G. crassaformis* include a shallow thermocline, high productivity in surface waters, and a distinct subsurface oxygen minimum where sinking organic matter is oxidized [10].

We selected about four specimens of *G. crassaformis* per samples in the 425-500 µm fraction from the core RC12-10 (23°N, 95°53 W, 3054 m) for stable isotope measurements (figure 4). Samples were analyzed on a MAT 253 mass spectrometer equipped with an automated Kiel Carbonate device at Georgia Institute of Technology. Long term reproducibility is < 0.07‰ for $\delta^{18}O$.

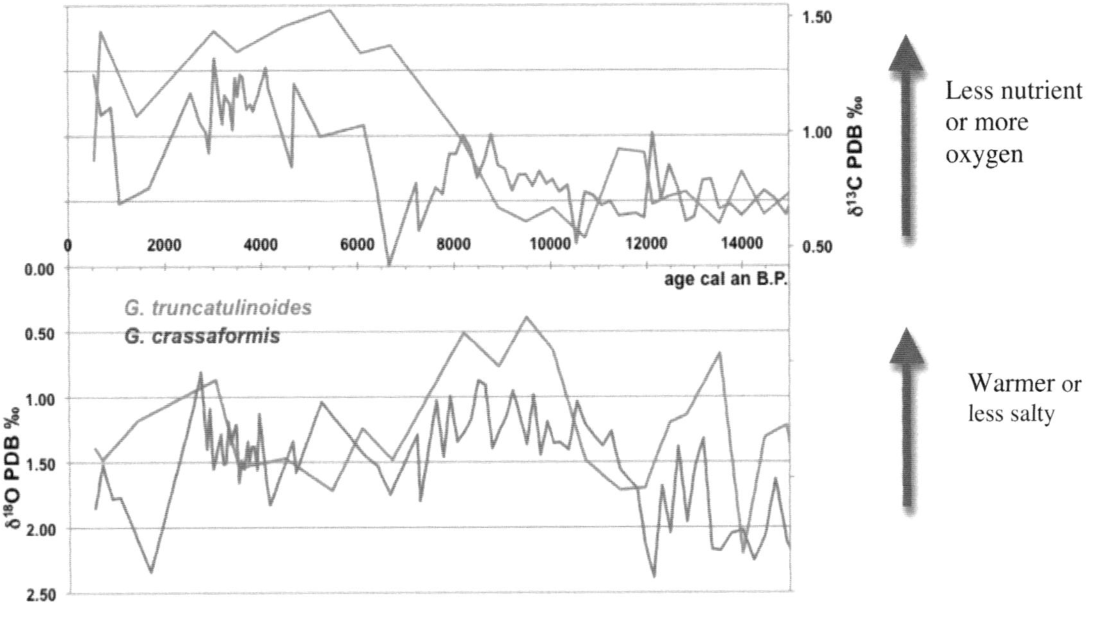

Figure 4: Stable isotopic analysis of *G. crassaformis* (blue) and *G. truncatulinoides* (orange) in the core RC12-10 in the Gulf of Mexico. $\delta^{13}C$ and $\delta^{18}O$ are respectively plotted on the upper and lower panel.

The isotopic composition of *G. crassaformis* in the core RC12-10 has been analyzed at a much higher resolution than for *G. truncatulinoides* which explain the higher variability. More importantly, *G. crassaformis* $\delta^{18}O$ signal doesn't show any large amplitude change over the last 11 ka. The large decrease in *G. truncatulinoides* $\delta^{18}O$ value between 8-10 ka, interpreted as an upward migration, is not seen in *G.*

crassaformis record. *G. crassaformis* abundance is very low at the top of the core, we therefore don't interpret the heavier value over the last 2 ka.

The $\delta^{13}C$ signal of both deep-dwelling planktonic species in the RC12-10 core is also plotted on figure 4. The $\delta^{13}C$ of deep-dwelling planktonic, less influenced by atmospheric exchange than surface dwelling foraminifera, should reflect the seawater biological productivity (organic matter formation/oxidation ratio) and thermocline ventilation. The *G. truncatulinoides* $\delta^{13}C$ values are low during the deglaciation/early Holocene and increase by about 0.7‰ around 8 ka. This increase reflects a more nutrient depleted or more oxygenated habitat which is at odds with a deeper calcification depth in the late Holocene. The mean $\delta^{13}C$ of *G. crassaformis* is also higher in the late than in the early Holocene but the amplitude of the change is about 0.4‰.

Another environmental piece of information can be retrieved from *G. crassaformis* abundance in the core RC12-10 [15]. This species was less abundant between 11-13 ka (age cal) and seems to almost disappear from the Gulf of Mexico around 8 ka (age cal) [11]. Knowing the ecological preference of this species, this probably reflects a drop in productivity and/or more oxygenated subsurface condition around 8 ka.

3. Discussion

3.1. Estimating physical water column properties changes

Calcification habitat of *G. truncatulinoides* in the Florida Straits shows a step-like cooling around 8 ka, contemporaneous with the large $\delta^{18}O$ signal increase previously identified. This cooling is expected with a migration toward deeper calcification depth in the late Holocene. The reconstructed seawater $\delta^{18}O$/salinity values don't show such transition between the early and the late Holocene.

Foraminifera, and particularly deep-dwelling species, add a secondary calcite (or crust) to their shelf when they sink at deeper level in the water column [12]. One could argue that different amount of crust in the test between the deglaciation and the Holocene might creates the observed signals. Several arguments oppose to calcite crust influence: 1) The *G. truncatulinoides* $\delta^{18}O$ signal at 8-10 ka is observed in three different cores where specimens were picked in different size fractions; 2) it would be odd if *G. truncatulinoides* added a crust to its test at the same time in the three different locations; 3) [12] showed that calcite crust has heavier $\delta^{18}O$ and $\delta^{13}C$. $\delta^{18}O$ values around 8-10 ka are depleted but not the $\delta^{13}C$ values (figure 4).

We use *G. crassaformis* that has, at present day, a similar calcification depth than *G. truncatulinoides* to check whether or not the shallowing of calcification depth was a common feature among thermocline dwelling foraminifera. The $\delta^{18}O$ of *G. crassaformis* don't show the large change seen in *G. truncatulinoides* signals. *G. crassaformis* probably kept the same calcification habitat over the last 11 ka and this habitat experienced no significant temperature or salinity variation. From temperature, seawater $\delta^{18}O$ reconstruction from *G. truncatulinoides* in the Florida Straits and *G. crassaformis* data from the Gulf of Mexico core, we therefore conclude that seawater temperature and salinity don't seem to be the main cause for the observed *G. truncatulinoides* migration.

3.2. Estimating nutrient/productivity water column changes

The $\delta^{13}C$ of *G. truncatulinoides* and abundance change of *G. crassaformis* in the Gulf of Mexico show large changes contemporary with the migration of *G. truncatulinoides*. The quasi-disappearance of *G. crassaformis* around 8 ka can be interpreted as a decrease in productivity and/or more oxygenated sub-surface water. This is In agreement with the $\delta^{13}C$ enrichment of *G. truncatulinoides* in RC12-10 from the early to the late Holocene. Over the Western Atlantic Carolina slope, [19] also found a more productive surface water from 15 to 10 ka than over the Holocene. These authors suggest a change in the mode of productivity that might have resulted from larger seasonality or winter mixing. The Cd/Ca data on the Bermuda Rise also indicates a decrease in nutrient content from about 10 ka to the Holocene [2, 3]. Change in surface water of the Gulf of Mexico occurs at this same period [15]. Although all these observations might have different origins, they tend toward a general change in productivity over the western subtropical Atlantic around 10-8 ka. Possible mechanism for the change in nutrient content/productivity can be ocean circulation change [2, 3] or the end of deglacial discharge over these continental slope areas. The input of meltwater, charged with nutrient and particles, in the Gulf of Mexico stopped at 9.5 ka [1]. We hypothesize that the migration of *G. truncatulinoides* calcification depth might have been triggered by change is nutrient and productivity. The lack of knowledge on *G. truncatulinoides* ecology prevents further interpretation but point out to important work to carry on.

4. Conclusion

With these new developments on subsurface water condition in the Gulf of Mexico and the Florida Straits, we show that:

- *G. truncatulinoides* was calcifying at warmer temperature when it recorded the heavy $\delta^{18}O$ which confirms a shallower habitat.
- Subsurface temperature and salinity probably didn't change drastically over the last 11 ka and don't explain the migration.
- The cause for *G. truncatulinoides* migration lies probably in a major nutrient/productivity change.

Further work is necessary to answer the questions:

- What *G. truncatulinoides* feed on?
- What caused this drastic change in water nutrient composition?
- How is it related to change in ocean circulation?

Acknowledgements

This work was supported by NSF grant OCE-0648258 and a grant from the Comer Science and Education Foundation. We are grateful to R. Z. Poore who provides samples of the core RC12-10 and M. W. Schmidt who performed *G. truncatulinoides* isotopic analyses on the core KNR166-2 29JPC. Many thanks to Tammy Chiang for help with sample preparation. CC participation to the YSM PAGES meeting in Corvalis 2009 was made possible by a U.S. National Science Foundation grant.

References

[[1] Aharon P 2003 Meltwater flooding events in the Gulf of Mexico revisited: Implication for rapid climate changes during the last deglaciation *Paleoceanography* **18** doi: 10.1029/2002PA000840

[2] Alley R B, Mayewski P A, Sowers M, Stuiver M, Taylor K C and Clark P U 1997 Holocene climatic instability: A prominent, widespread event 8200 years ago *Geology* **25** 483-6

[3] Boyle E and Keigwin L D 1987 North Atlantic thermohaline circulation during the past 20,000 years linked to high-latitude surface temperature *Nature* **330** 35-40

[4] Cléroux C, Cortijo E, Anand P, Labeyrie L, Bassinot F, Caillon N and Duplessy J C 2008 Mg/Ca and Sr/Ca ratios in planktonic foraminifera: proxies for upper water column temperature reconstruction *Paleoceanography* doi:10.1029/2007PA001505

[5] Cléroux C, Cortijo E, Duplessy J C and Zahn R 2007 Deep-dwelling foraminifera as thermocline temperature recorders *Geochemistry Geophysics Geosystems* **8** 10.1029/2006GC001474

[6] Cléroux C, Lynch-Stieglitz J, Schmidt M W, Cortijo E and Duplessy J-C 2009 Evidence for calcification depth change of Globorotalia truncatulinoides between deglaciation and Holocene in the Western Atlantic Ocean *Marine Micropaleontology* **73** 57-61

[7] Deuser W G and Ross E H 1989 Seasonally abundant planktonic foraminifera of the Sargasso Sea - succession, deep-water fluxes, isotopic composition and paleoceanographic implication *Journal of Foraminifera Research* **19** 268-93

[8] Farmer E C, Kaplan A, deMenocal P B and Lynch-Steiglitz J 2007 Corroborating ecological depth preferences of planktonic foraminifera in the tropical Atlantic with the stable oxygen isotope

ratios of core top specimens *Paleoceanography* **22** doi 10.1029/2006PA001361

[9] Greaves M, Caillon N, Rebaubier H and group a i 2008 Interlaboratory comparison study of calibration standards for foraminiferal Mg/Ca thermometry *Geochemistry Geophysics Geosystems* **9** doi 10.1029/2008GC001974

[10] Jones J I 1967 Significance of the distribution of planktonic foraminifera in the equatorial Atlantic undercurrent *Micropaleontology* **13** 489-501

[11] LoDico J M, Flower B P and Quinn T M 2006 Subcentennial-scale climatic and hydrologic variability in the Gulf of Mexico during the early Holocene *Paleoceanography* **21** 10.1029/2005PA001243

[12] Lohmann G 1995 A modek for variation in the chemistry of planktonic foraminifera due to secondary calcification and selective dissolution *Paleoceanography* **10** 445-57

[13] Lynch-Stieglitz J, Curry W B C and Lund D C 2009 Florida Straits Density Structure and Transport over the last 8,000 Years *Paleoceanography* doi 10.1029/2008PA001717

[14] Mulitza S, Dürkoop A, Hale W, Wefer G and Niebler H S 1997 Planktonic foraminifera as recorders of past surface-water stratification *Geology* **25** 335-8

[15] Poore R Z, Dowsett H J, Verardo S and Quinn T M 2003 Millennial- to century-scale variability in Gulf of Mexico Holocene climate records *Paleoceanography* **18** doi 10.1029/PA000868

[16] Shackleton N J 1974 Attainment of isotopic equilibrium between ocean water and benthonic foraminifera genus Uvigerina : isotopic changes in the ocean during the last glacial. In: *Les méthodes quantitatives d'étude des variations du climat au cours du Pleistocène,* (Gif sur Yvette: CNRS) pp 203-9

[17] Tedesco K A, Thunell R C, Astor Y and Muller-Karger F 2007 The oxygen isotope composition of planktonic foraminifera from the cariaco Basin, Venezuela: Seasonal and interannual variations *Marine Micropaleontology* **62** 180-93

[18] Waelbroeck C, Labeyrie L, Michel E, Duplessy J C, McManus J F, Lambeck K, Balbon E and Labracherie M 2002 Sea-level and deep water temperature changes derived from benthic foraminifera isotopic records *Quaternary Science Reviews* **21** 295-305

[19] Yasuhara M, Cronin T M, deMenocal P B, Okahashi H and Linsley B K 2008 Abrupt climate change and collapse of deep-sea ecosytems *PNAS* **105** 1556-60

Assessing the potential of Southern Caribbean corals for reconstructions of Holocene temperature variability

C Giry[1], T Felis[1], S Scheffers[2] and C Fensterer[3]

[1] MARUM - Center for Marine Environmental Sciences, University of Bremen, 28359 Bremen, Germany
[2] School of Environmental Science and Management, Southern Cross University, PO 147 Lismore, Australia.
[3] Heidelberger Akademie der Wissenschaften, Forschungsstelle Radiometrie, Heidelberg, Germany.

E-mail: cgiry@marum.de

Abstract. We present a 40-year long monthly resolved Sr/Ca record from a fossil *Diploria strigosa* coral from Bonaire (Southern Caribbean Sea) dated with U/Th at 2.35 ka before present (BP). Secondary modifiers of this sea surface temperature (SST) proxy in annually-banded corals such as diagenetic alteration of the skeleton and skeletal growth-rate are investigated. Extensive diagenetic investigations reveal that this fossil coral skeleton is pristine which is further supported by clear annual cycles in the coral Sr/Ca record. No significant correlation between annual growth rate and Sr/Ca is observed, suggesting that the Sr/Ca record is not affected by coral growth. Therefore, we conclude that the observed interannual Sr/Ca variability was influenced by ambient SST variability. Spectral analysis of the annual mean Sr/Ca record reveals a dominant frequency centred at 6-7 years that is not associated with changes of the annual growth rate. The first monthly resolved coral Sr/Ca record from the Southern Caribbean Sea for preindustrial time suggests that fossil corals from Bonaire are suitable tools for reconstructing past SST variability. Coastal deposits on Bonaire provide abundant fossil *D. strigosa* colonies of Holocene age that can be accurately dated and used to reconstruct climate variability. Comparisons of long monthly resolved Sr/Ca records from multiple fossil corals will provide a mean to estimate seasonality and interannual to interdecadal SST variability of the Southern Caribbean Sea during the Holocene.

1. Introduction

During their growth, massive scleractinian corals living in shallow water of tropical and subtropical oceans incorporate a large array of geochemical tracers in their annually-banded aragonite skeleton that can be measured to retrieve environmental parameters [1, 2]. Robust chronological control allows coral skeleton microsampling at subseasonal resolution. Key oceanographic parameters such as sea surface temperature (SST) and sea surface salinity inferred from long-lived annually-banded coral colonies have recently improved our understanding of the tropical and subtropical climate system [3-10]. The strontium to calcium ratio (Sr/Ca) in coral skeletons [11, 12] has been proven to be the most reliable salinity-independent SST proxy. In recent decades, the inverse relationship between SST and coral Sr/Ca has been largely used to evaluate past SST variability of tropical and subtropical oceans [13-15]. To date, establishment of long coral-based SST reconstructions of the Western North Atlantic and Caribbean that extend beyond sparse and short instrumental data are underrepresented. Exceptions include the studies by Winter *et al.* (2000) [16], Kuhnert *et al.* (2002) [17], Winter *et al.* (2003) [18], Kuhnert *et al.* (2005) [19], Kilbourne *et al.* (2008) [20], Goodkin *et al.* (2008) [21] and Saenger *et al.* (2009) [22].

Paleoclimate reconstructions of the tropical Atlantic inferred from sediment cores and speleothems have provided evidence for variable hydrological conditions over the last thousands of years [23-25]. On orbital timescales, Southern Caribbean rainfall changes are linked to the mean position of the Intertropical Convergence Zone (ITCZ) which is controlled by a meridional SST gradient in the equatorial sector. On shorter timescales such as interannual to interdecadal, instrumental observations suggest that the dynamics of the ITCZ and thus, regional SSTs, are influenced by natural modes of ocean/atmosphere variability of specific periodicity such as the El Niño-Southern Oscillation (ENSO), the North Atlantic Oscillation (NAO) and the Atlantic Multidecadal Oscillation (AMO) emanating from tropical and subtropical oceans [26, 27]. The competing influence of these modes of climate variability on tropical Atlantic climate is called the Tropical Atlantic Variability (TAV) [26].

In the Southern Caribbean Sea, a subdecadally-resolved bulk-Ti content record from anoxic sediment of the Cariaco Basin shows that rainfall changes over northern South America are influenced, on orbital timescales, by the mean latitudinal position of the ITCZ [24]. Although this runoff proxy indicates long-term changes of local precipitation toward drier conditions driven by insolation decrease over the Northern Hemisphere during the Holocene, this record suggests that Southern Caribbean climate has encountered periods of highly-variable hydrological conditions with high-amplitude and high-frequency variability of regional rainfall [23, 24]. Whether those abrupt hydrological shifts are associated with changes in the magnitude of teleconnected modes of climate variability emanating from the Pacific or the Atlantic oceans still remains unclear.

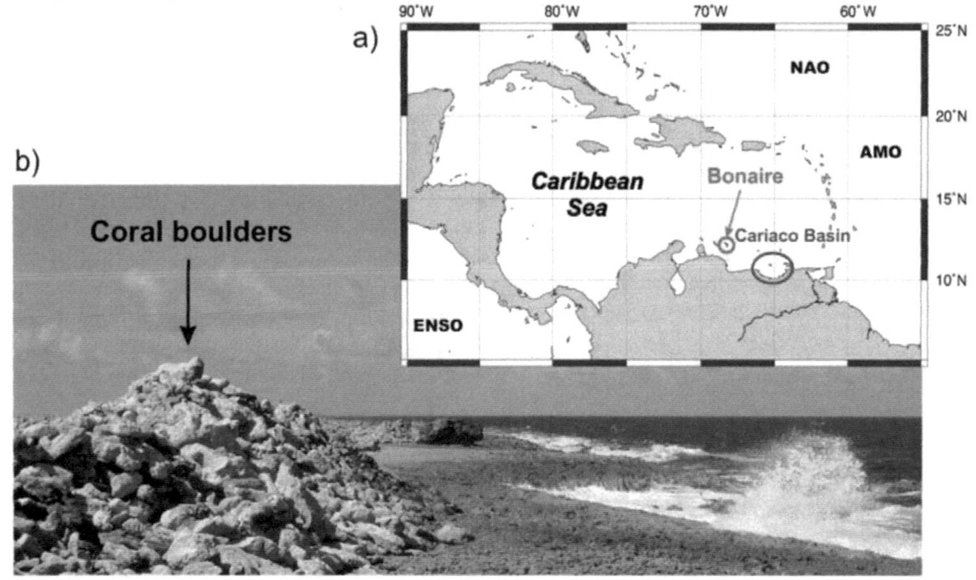

Figure 1. a) Bonaire is an open-ocean island located in the southern Caribbean Sea where local sea surface temperature variability is influenced by natural modes of ocean/atmosphere variability emanating from tropical and subtropical oceans, such as the El Niño-Southern Oscillation (ENSO), the North Atlantic Oscillation (NAO) and the Atlantic Multidecadal Oscillation (AMO). b) Accumulations of Holocene coral boulders transported by extreme waves (storm-induced and/or tsunamis-induced) form ramparts up to 10 m above sea level on Bonaire [28].

In this study, we use a well-dated and well-preserved fossil coral from Bonaire to reconstruct seasonal and interannual SST variability during a time window of the late Holocene. Coral Sr/Ca in *Diploria strigosa* from Guadeloupe has been recently calibrated against SST and has proven capabilities for reconstructing interannual and interdecadal SST variability in the Caribbean [29]. Hetzinger *et al.*, (2006) found a monthly mean Sr/Ca-SST relationship of -0.041 mmol/mol per °C with instrumental SST data [30]. Secondary modifiers of this SST proxy such as diagenetic alteration and skeletal growth-rate are investigated. Finally, the dominant frequency of SST variability documented in the coral Sr/Ca record is discussed.

2. Material and Methods

2.1 Regional setting

Bonaire (Netherlands Antilles) is a small island in the Southern Caribbean Sea located ~ 100 km north of the Venezuelan coast (12°10'N; 68°18'W) and ~400 km away from Cariaco Basin (figure 1a). This open-ocean island lies at the southern edge of the hurricane belt [31, 32] and is partially fringed by coral reefs since the mid Holocene [28, 33]. Tropical storms and hurricanes passing north of Bonaire create intense wave action that cause damage to the reef and significant shoreline modifications [34, 35]. Coastal deposits commonly found on Bonaire consist of decimetric to metric coral rubbles and boulders derived from adjacent coral reef (figure 1b). The modern climate of Bonaire is tropical-arid characterised by low annual rainfall (~ 500 mm/year), high evaporation rate due to year-long easterly trade winds and high annual-mean SST (~26.8 °C) ranging from 25.7 °C (February) to 27.9 °C (October) [30].

2.2 Coral sampling and ^{230}Th/U-dating

Hurricane and/or tsunami deposits on Bonaire provide numerous fossil corals of Holocene age. A 40-cm-long core has been drilled from a fossil *Diploria strigosa* coral colony (BON-6-A) from one of these deposits at Washikemba (12°8'31.26"N, 68°11'53.10"W) on the windward coast of Bonaire (figure 1b). ^{230}Th/U-dating at the Heidelberg Academy of Sciences indicates a coral age of 2.406 ± 0.049 ka (2σ-uncertainty of 49 years), which is equivalent to 2.347 ± 0.049 ka before present (BP, where present is defined as AD 1950). Analyses were performed with a Finnigan Thermal Ionization Mass Spectrometer (TIMS) MAT 262 RPQ. Sample preparation and analytical details are similar as described in Scholz *et al.* (2004) [36]. Calibration of the added U and Th spike solutions is described in Hoffmann *et al.* (2007) [37]. Ages were calculated using the half-lives of Cheng *et al.* (2000) [38]. All ages were corrected for the effect of detrital contamination using a bulk earth ^{232}Th/^{238}U weight ratio of 3.8 and assuming secular equilibrium between ^{238}U, ^{234}U and ^{230}Th. The effect of this correction, however, is 1 year and, thus, within the range of 2σ-uncertainty of 49 years and negligible. The reliability of the determined ^{230}Th/U-age was checked using established criteria, such as the initial (^{234}U/^{238}U) ratio in agreement with the (^{234}U/^{238}U) ratio of modern seawater (1.146, [39]) and U concentrations comparable to modern corals of the same species. For the studied fossil coral all these criteria are fulfilled. Thus, its age is considered as reliable.

2.3 Diagenetic investigations

The coral core was sectioned into 7 mm-thick slabs and investigated for possible diagenetic alteration of the skeleton using different methods. The slabs were X-rayed in order to identify both the annual density-banding pattern and areas of anomalous density patches potentially reflecting diagenetic alterations such as calcite or secondary aragonite infillings. In addition, for investigating diagenetic alteration that is representative of the entire coral slab, samples for powder X-ray diffraction (XRD), Scanning Electron Microscope (SEM) imaging and thin-section microscopy analyses were extracted from the middle of the slab. XRD was applied in order quantify the calcite content of the sample. As crystals of inorganic aragonite cannot be detected with XRD, we used SEM and thin-section microscopy to identify the presence of secondary aragonite overgrowth that can potentially bias Sr/Ca-based SST reconstructions.

2.4 Microsampling

Based on the annual density-banding pattern of the coral skeleton observed on X-radiographs, we targeted a sampling resolution of about 12 samples per year, allowing an evaluation of proxy variability at monthly resolution. For each sample, an average of 1 mg skeletal powder was drilled along the centre of the solid and dense theca wall, using a 0.6 mm diameter drill bit and applying a sampling depth of 1.5-2 mm and an average sampling increment of 0.55 mm.

2.5 Sr/Ca analyses

For Sr/Ca analyses 0.25-0.30 mg sample powder was dissolved in 7 mL 2 % suprapure HNO_3 containing 1 ppm Sc (Scandium) as internal standard. Measurements were performed on a Perkin-Elmer Optima 3300R simultaneous radial ICP-OES using a CETAC U5000-AT ultrasonic nebulizer (University of Bremen). Element wavelengths were detected simultaneously in 3 replicates (Ca 317.933 nm, Ca 422.673 nm, Sr 421.552 nm, Sc 361.383 nm and Mg 280.271 nm). Intensities of the two Ca lines were averaged to compensate for possible plasma-induced variations in signal intensity. Measurements of a laboratory coral standard after each sample allowed offline correction for instrumental drift. Relative standard deviation of the Sr/Ca determinations was better than 0.2 %.

2.6 Coral chronology

An average of 11.8 samples per year ranging from 8 to 14 has been extracted from the dense theca wall over 40 couplets of high- and low-density bands. The age model for the coral chronology is based on the seasonal banding pattern seen in the X-radiographs (figure 2a) and the annual cycles in the skeletal Sr/Ca record. We assume that no significant changes the timing of the seasonal maximum and minimum of SST occurred during the late Holocene. Therefore, maximum and minimum Sr/Ca values in any given year were assigned to the coldest (February) and warmest month (October) of present-day conditions, and subsequently interpolated to a monthly resolution following established procedures [6, 7, 15]. The skeletal annual extension rate was calculated by using the distance from a maximum Sr/Ca value in a given year to the maximum value of the following year and annual mean Sr/Ca was calculated by averaging monthly data of each year.

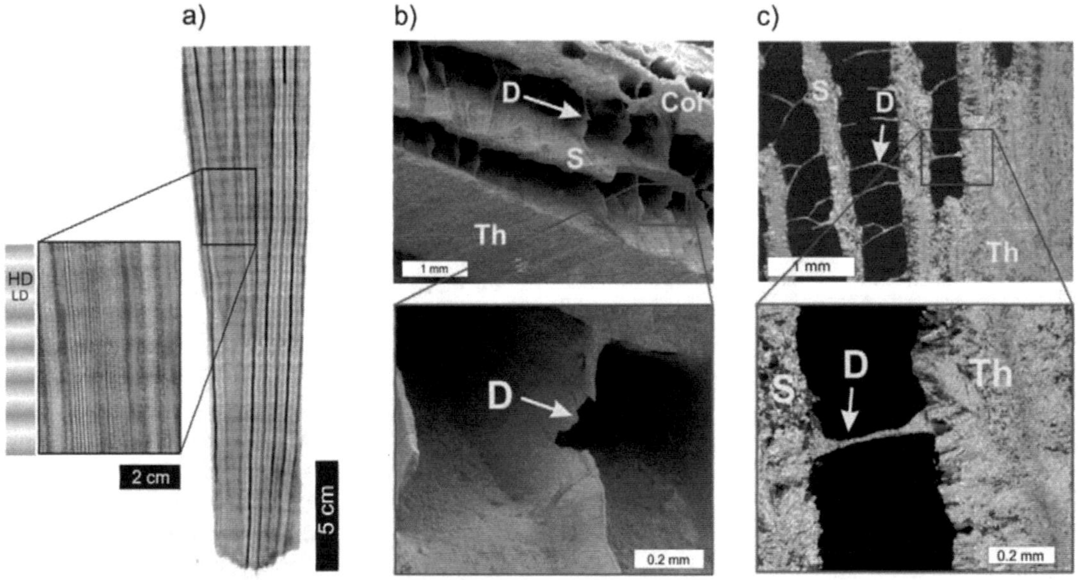

Figure 2. Investigation of the skeleton of the 2.35 ka BP Bonaire coral for possible diagenetic alteration. a) X-radiograph positive print of the coral slab (position of microsampling paths indicated by red lines) shows that powdered microsamples were extracted along the dense theca wall over annual couplets of high- and low-density bands (HD and LD, respectively). b) Scanning electron microscope images and c) thin section photomicrographs performed on samples from the centre of the slab reveal that individual skeletal elements such as theca (Th), septa (S), columella (Col) and dissepiments (D) have smooth surfaces with no indication for overgrowth by diagenetic cements, resulting in an excellent preservation of primary porosity of the coral skeleton.

3. Results and discussion

3.1 Preservation of coral skeleton

Diagenetic alterations such as dissolution, secondary precipitation of inorganic aragonite and calcite are known to influence the Sr/Ca ratio of coral skeletons [40-43]. XRD reveals a calcite content of the skeleton of <0.5 %. X-radiographs of the coral slab do not reveal anomalous density patches that are typical for diagenetic textures (figure 2a). SEM analyses show a smooth surface of individual skeletal elements without any overgrowth or dissolution patterns (figure 2b). Thin sections reveal that biogenic aragonite crystals are not overgrown by diagenetic cements and that centres of calcification are not dissolved (figure 2c). In summary, diagenetic investigation of the 2.35 ka BP coral reveals that the skeleton is pristine. Therefore, we conclude that the Sr/Ca variability discussed in the next section is not affected by diagenetic alteration.

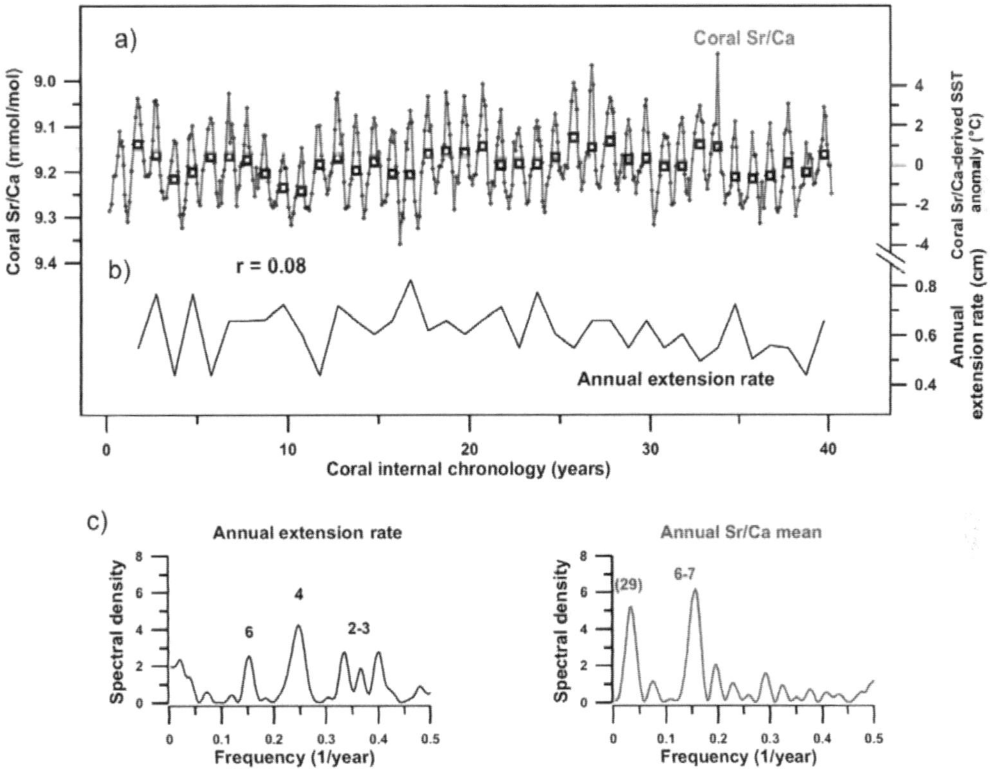

Figure 3. a) A 40-year long monthly resolved Sr/Ca record from a 2.35 ka BP coral from Bonaire (red line) reveals clear annual cycles and interannual variability. The corresponding annual mean record is also shown (open squares). Coral Sr/Ca-derived sea surface temperature anomaly is calculated by using the regression slope of -0.041 mmol/mol per °C from Hetzinger *et al.* (2006) [29]. The non-significant correlation between annual extension rate b) and the corresponding annual mean Sr/Ca record (r = 0.08) suggests that coral growth is not affecting skeletal Sr/Ca on interannual and longer timescales. c) Spectral analyses of the records of annual extension rate and annual mean Sr/Ca reveal that these two parameters seem to be influenced by distinct forcing. Prominent periodicities are indicated (years).

3.2 Coral Sr/Ca record

The 40-year long monthly resolved record shows clear annual cycles in Sr/Ca with average amplitude of 0.206 mmol/mol and displays interannual variability (figure 3a). The Sr/Ca interannual variability calculated by the standard deviation of the annual mean Sr/Ca record is 0.028 mmol/mol. Using a relationship of -0.041 mmol/mol per °C, based on a monthly coral Sr/Ca-SST regression of a *Diploria strigosa* from Guadeloupe [29], our coral record suggests that the average amplitude of the annual cycle of SST at 2.35 ka BP was 4.9 °C with an estimated interannual variability of 0.7 °C. However, using the commonly used relationship between coral Sr/Ca and SST of -0.06 mmol/mol per °C [44], our proxy record indicates an average seasonality and interannual variability of 3.4 °C and 0.5 °C, respectively.

Observations of instrumental SST around Bonaire taken from the SODA reanalysis data set (gridbox centred at 68.25°W; 12.25°N) [30] for a 40-year period (1960-2000) reveals that the modern amplitude of the annual cycle and interannual variability is about 2-3 °C and 0.35 °C, respectively. These observations suggest that slightly higher SST seasonality and interannual variability compared to today are recorded in the Sr/Ca of the 2.35 ka BP coral.

Recent understanding in the calcification and architecture of the skeleton of brain corals revealed some problems related to microsampling such corals. Time-transgressive problems associated with microsampling the convoluted walls of slow-growing *Diploria* sp. and thus, integration of skeletal material precipitated at different months are known to produce a dampened seasonal cycle [45]. Although our record is retrieved from a similar convoluted skeletal element of a *Diploria strigosa* coral growing at a rate of 0.62 cm per year, such problems can not explain the higher amplitude of the annual cycle of SST relative to today reconstructed from the 2.35 ka BP coral. Incorporation of strontium into the skeleton of *Diploria* sp. has been shown to be partially related to the annual extension rate, where annual mean Sr/Ca in some colonies is negatively correlated with annual growth-rate [46, 47]. In our study, the annual skeletal extension rate was calculated as the distance from a maximum Sr/Ca value in a given year to the maximum value of the following year. The non-significant correlation ($r = 0.08$) between annual extension rate and the corresponding annual mean Sr/Ca suggests that extension rate is not affecting the skeletal Sr/Ca signal on interannual and longer timescales. Moreover, to ensure that there is in fact no influence of growth rate on the Sr/Ca signal, we further investigate the relationship between seasonal Sr/Ca and growth. Work by Goodkin *et al.* (2005) [47] found differences in the winter and summer season calibrations to SST attributed to seasonal variations of extension rate. This study examined the relationship between Sr/Ca and annual growth rate and they found that the summer Sr/Ca correlates strongly to annual skeletal extension rate whereas winter Sr/Ca showed no correlation to growth rate. Their observation suggests that subannual changes in growth rate affect the Sr/Ca signal. Although a potential relationship between Sr/Ca and SST on seasonal timescales can not be investigated in our fossil coral, we found that seasonal Sr/Ca values for winter, spring, summer and fall do not show any significant correlations with annual extension rate (not shown). Therefore, we assume that a seasonal growth dependence of Sr/Ca observed in slow-growing *Diploria labyrinthiformis* corals from Bermuda [45, 47] is not affecting the Sr/Ca in our fossil *Diploria strigosa* coral from Bonaire.

3.3 Southern Caribbean interannual SST variability at 2.35 ka BP

Spectral analysis of the annual mean Sr/Ca record exhibits a dominant periodicity of 6-7 years (figure 3c). Another spectral peak centred around 29 years is beyond the limit of low-frequency variability that can be inferred from the data, given the length of the time series (40 years). Spectral analysis of the annual growth-rate record indicates a dominant periodicity of 4 years which is not seen in the spectral analysis of the Sr/Ca record (figure 3c). The discrepancies in the results of spectral analysis between annual growth rate and annual mean Sr/Ca further strengthen our conclusion that the Sr/Ca signal on interannual and longer timescales is not affected by growth rate, and therefore reflects past SST variability.

Instrumental observations of Southern Caribbean climate have demonstrated that regional SST variations are linked to the strength and direction of trade winds which also control the regional evaporation/precipitation ratio [48]. The atmospheric circulation over the Southern Caribbean is controlled by the competing influence of modes of climate variability such as ENSO, NAO and AMO that control the magnitude of SST variability in tropical Atlantic sector [26]. Modern coral records from western Atlantic Ocean have demonstrated the non-stationary features of low-latitude Atlantic multidecadal climate variability [49]. Moreover, instrumental observations reveal that the interannual component of SST variability is non-stationary making accurate climate forecasts more difficult to achieve [50]. In this context, our preliminary finding that interannual variability in Southern Caribbean SST at 2.35 ka BP was dominated by a 6-7 year periodicity, which is not observed in our spectral analysis of regional SST [30] (not shown), support the non-steady behaviour of tropical Atlantic climate during the late Holocene. Future work on extending this Sr/Ca record over the entire growth period of the coral (>60 years) will provide a more robust estimate of interannual variations in Southern Caribbean SST during the late Holocene.

4. Conclusions

The present study investigates the potential of a fossil *Diploria strigosa* coral from Bonaire for reconstructing Southern Caribbean SST variability during the Holocene. The results suggest that fossil corals from this open-ocean island provide a unique opportunity for reconstructing interannual SST variability because of their well-preserved skeleton and extension-rate independent skeletal Sr/Ca ratio on interannual and longer timescales. Our record provides the first monthly resolved Sr/Ca-based SST reconstruction of the Southern Caribbean Sea for preindustrial time and suggests that annually-banded fossil corals from Bonaire can provide seasonal and interannual oceanographic information of the Southern Caribbean sea surface during the Holocene. Furthermore, future work on paired analyses of Sr/Ca and $\delta^{18}O$ of Holocene corals will provide new insights into both SST and hydrology variability of the Southern Caribbean Sea on seasonal and interannual timescales, e.g., during intervals when the nearby Cariaco Basin sediments document highly variable hydrological conditions over northern South America. Finally, monthly-resolved Sr/Ca and $\delta^{18}O$ coral records from Bonaire are expected to shed more light on Pacific versus Atlantic and tropical versus extratropical controls on Southern Caribbean climate during the Holocene.

Acknowledgments

We thank the Government of the Island Territory of Bonaire (Netherlands Antilles) for research and fieldwork permission, Elsmarie Beukenboom (STINAPA Bonaire National Parks Foundation) for support, J. Pätzold for support with coral drilling and discussion, M. Kölling and S. Pape for Sr/Ca analyses, K.-H. Baumann for SEM analyses, D. Scholz, H. Kuhnert and M. McCulloch for discussion and two anonymous reviewers for valuable suggestions that greatly improved this manuscript. This work was funded by Deutsche Forschungsgemeinschaft (DFG) under the Special Priority Programme INTERDYNAMIK, through a grant to T.F. (FE 615/3-1; CaribClim). C.F. was supported by DFG (MA 821/37-1; CaribClim). C.G. acknowledges INTERDYNAMIK (DFG) for a travel grant to attend the PAGES YSM/OSM and additional support by GLOMAR (Bremen International Graduate School for Marine Sciences) that is funded by DFG.

References

[1] Felis T and Pätzold J, ed. *Climate records from corals*. Marine Science Frontiers for Europe, ed. G. Wefer, F. Lamy and Mantoura F. 2003, Springer-Verlag: Berlin. 11-27.
[2] Gagan M K, Ayliffe L K, Beck J W, Cole J E, Druffel E R M, Dunbar R B, and Schrag D P, 2000 *Quaternary Science Reviews*, **19** 45-64
[3] Abram N J, Gagan M K, Cole J E, Hantoro W S, and Mudelsee M, 2008 *Nature Geosci*, **1** 849-853
[4] Cobb K M, Charles C D, Cheng H, and Edwards R L, 2003 *Nature*, **424** 271-276
[5] Cole J E, Fairbanks R G, and Shen G T, 1993 *Science*, **260** 1790-1793

[6] Felis T, Lohmann G, Kuhnert H, Lorenz S J, Scholz D, Patzold J, Al-Rousan S A, and Al-Moghrabi S M, 2004 *Nature*, **429** 164-168

[7] Felis T, Pätzold J, Loya Y, Fine M, Nawar A H, and Wefer G, 2000 *Paleoceanography*, **15** 679-694

[8] Linsley B K, Kaplan A, Gouriou Y, Salinger J, deMenocal P B, Wellington G M, and Howe S S, 2006 *Geochem. Geophys. Geosyst.*, **7** Q05003, doi:10.1029/2005GC001115

[9] Tudhope A W, Chilcott C P, McCulloch M T, Cook E R, Chappell J, Ellam R M, Lea D W, Lough J M, and Shimmield G B, 2001 *Science*, **291** 1511-1517

[10] Hetzinger S, Pfeiffer M, Dullo W-C, Keenlyside N, Latif M, and Zinke J, 2008 *Geology*, **36** 11-14

[11] Beck J W, Edwards R L, Ito E, Taylor F W, Recy J, Rougerie F, Joannot P, and Henin C, 1992 *Science*, **257** 644-647

[12] Smith S V, Buddemeier R W, Redalje R C, and Houck J E, 1979 *Science*, **204** 404-407

[13] Asami R, Felis T, Deschamps P, Hanawa K, Iryu Y, Bard E, Durand N, and Murayama M, 2009 *Earth and Planetary Science Letters*, **288** 96-107

[14] Correge T, Gagan M K, Beck J W, Burr G S, Cabioch G, and Le Cornec F, 2004 *Nature*, **428** 927-929

[15] Felis T, Suzuki A, Kuhnert H, Dima M, Lohmann G, and Kawahata H, 2009 *Geology*, **37** 527-530

[16] Winter A, Ishioroshi H, Watanabe T, Oba T, and Christy J, 2000 *Geophys. Res. Lett.*, **27** 3365-3368

[17] Kuhnert H, Pätzold J, Schnetger B, and Wefer G, 2002 *Palaeogeography, Palaeoclimatology, Palaeoecology*, **179** 159-171

[18] Winter A, Paul A, Nyberg J, Oba T, Lundberg J, Schrag D, and Taggart B, 2003 *Geophys. Res. Lett.*, **30** 1163, doi:10.1029/2002GL016275

[19] Kuhnert H, Crüger T, and Pätzold J, 2005 *Geochem. Geophys. Geosyst.*, **6** Q04004, doi:10.1029/2004GC000786

[20] Kilbourne K H, Quinn T M, Webb R, Guilderson T, Nyberg J, and Winter A, 2008 *Paleoceanography*, **23** PA3220, doi:10.1029/2008PA001598

[21] Goodkin N F, Hughen K A, Doney S C, and Curry W B, 2008 *Nature Geosci*, **1** 844-848

[22] Saenger C, Cohen A L, Oppo D W, and Hubbard D, 2008 *Paleoceanography*, **23** PA3102, doi:10.1029/2007PA001572

[23] Cruz F W, Vuille M, Burns S J, Wang X, Cheng H, Werner M, Lawrence Edwards R, Karmann I, Auler A S, and Nguyen H, 2009 *Nature Geosci*, **2** 210-214

[24] Haug G H, Hughen K A, Sigman D M, Peterson L C, and Rohl U, 2001 *Science*, **293** 1304-1308

[25] Hodell D A, Brenner M, and Curtis J H, 2005 *Quaternary Science Reviews*, **24** 1413-1427

[26] Hurrell J W, Visbeck M, Busalacchi A, Clarke R A, Delworth T L, Dickson R R, Johns W E, Koltermann K P, Kushnir Y, Marshall D, Mauritzen C, McCartney M S, Piola A, Reason C, Reverdin G, Schott F, Sutton R, Wainer I, and Wright D, 2006 *Journal of Climate*, **19** 5100-5121

[27] Marshall J, Yochanan K, David B, Ping C, Arnaud C, Robert D, James H, Michael M, Saravanan R, and Martin V, *North Atlantic climate variability: phenomena, impacts and mechanisms*. 2001. p. 1863-1898.

[28] Scheffers A, 2004 *Quaternary International*, **120** 163-172

[29] Hetzinger S, Pfeiffer M, Dullo W-C, Ruprecht E, and Garbe-Schönberg D, 2006 *Geochem. Geophys. Geosyst.*, **7** Q10002, doi:10.1029/2006GC001347
[30] Carton J A and Giese B S, 2008 *Monthly Weather Review*, **136** 2999-3017
[31] Bries J M, Debrot A O, and Meyer D L, 2004 *Coral Reefs*, **23** 297-307
[32] Meyer D L, Bries J M, Greenstein B J, and Debrot A O, 2003 *Lethaia*, **36** 273-286
[33] Focke J W, 1978 *Marine Geology*, **28** M31-M41
[34] Scheffers A and Scheffers S, 2006 *Journal of Coastal Research*, **22** 1437-1450
[35] Scheffers S R, Haviser J, Browne T, and Scheffers A, 2009 *Quaternary International*, **195** 69-87
[36] Scholz D, Mangini A, and Felis T, 2004 *Earth and Planetary Science Letters*, **218** 163-178
[37] Hoffmann D L, Prytulak J, Richards D A, Elliott T, Coath C D, Smart P L, and Scholz D, 2007 *International Journal of Mass Spectrometry*, **264** 97-109
[38] Cheng H, Edwards R L, Hoff J, Gallup C D, Richards D A, and Asmerom Y, 2000 *Chemical Geology*, **169** 17-33
[39] Robinson L F, Belshaw N S, and Henderson G M, 2004 *Geochimica et Cosmochimica Acta*, **68** 1777-1789
[40] Cohen A L and Hart S R, 2004 *Paleoceanography*, **19** PA4031, doi:10.1029/2004PA001084
[41] Hendy E J, Gagan M K, Lough J M, McCulloch M, and deMenocal P B, 2007 *Paleoceanography*, **22** PA4101, doi:10.1029/2007PA001462
[42] McGregor H V and Abram N J, 2008 *Geochem. Geophys. Geosyst.*, **9** Q10013, doi:10.1029/2008GC002093
[43] McGregor H V and Gagan M K, 2003 *Geochimica et Cosmochimica Acta*, **67** 2147-2156
[44] Corrège T, 2006 *Palaeogeography, Palaeoclimatology, Palaeoecology*, **232** 408-428
[45] Cohen A L, Smith S R, McCartney M S, and Etten J v, 2004 *Mar. Ecol. Prog. Ser.*, **271** 147-158
[46] Goodkin N F, Hughen K A, and Cohen A L, 2007 *Paleoceanography*, **22** PA1214, doi:10.1029/2006PA001312
[47] Goodkin N F, Hughen K A, Cohen A L, and Smith S R, 2005 *Paleoceanography*, **20** PA4016, doi:10.1029/2005PA001140
[48] Giannini A, Kushnir Y, and Cane M A, 2000 *Journal of Climate*, **13** 297-311
[49] Saenger C, Cohen A L, Oppo D W, Halley R B, and Carilli J E, 2009 *Nature Geosci*, **2** 492-495
[50] Sutton R T and Hodson D L R, 2003 *Journal of Climate*, **16** 3296-3313

IOP Conf. Series: Earth and Environmental Science **9** (2010) 012022 doi:10.1088/1755-1315/9/1/012022

A forward model of cave dripwater $\delta^{18}O$ and application to speleothem records

Sarah A Truebe, Toby R Ault and Julia E Cole

Department of Geosciences, University of Arizona, 1040 E. 4th St, Tucson, AZ, 85721, USA

E-mail: struebe@email.arizona.edu

Abstract. Speleothem oxygen isotope records in arid regions are typically interpreted as indicators of the total precipitation amount and/or its seasonal balance. Such studies rarely address the potential influence of groundwater mixing processes on $\delta^{18}O$ variability of cave dripwater. Here, we develop a model of oxygen isotopes in dripwater, which we compare to water and stalagmite measurements from Cave of the Bells, Arizona. We simulate moisture flux from surface to cave as a two-layer "leaky-bucket" model. In observations and the model, fed with modern climate data, dripwater $\delta^{18}O$ is most comparable to that of winter precipitation. We show that seasonality and duration of the regional summer monsoon affect how much summer precipitation reaches the cave. We employ a Monte Carlo method to specify statistically realistic ranges for input climate variables and produce time series and variance spectra of cave drips. The spectra of our synthetic $\delta^{18}O$ series exhibit a high degree of variance at decadal to multidecadal frequencies, despite being driven by synthetic data that includes only a seasonal cycle. This suggests that some background level of variance in speleothem $\delta^{18}O$ records could be due to nonclimatic processes, such as subsurface water storage and mixing. Interpreting climatic vs. nonclimatic controls on speleothem $\delta^{18}O$ variance could be achieved by replicating records from different caves.

1. Introduction

Population in the Southwestern United States is growing rapidly, and in this already water-stressed region, future changes in water availability are critical to sustainable occupation. The arid Southwest receives rainfall from two seasonally distinct types of systems: winter frontal storms (October-March) and summer monsoon thunderstorms (mid-June-September) each contribute half of the annual rainfall total (~400mm) in southern Arizona [1,2]. Climate model studies suggest that winter precipitation will decrease as global temperatures rise [3], but the future of the monsoon remains poorly constrained. This uncertainty arises in part because global climate models do not simulate monsoonal precipitation realistically [4], and because monsoon rainfall is inherently noisy; variance about the precipitation amount mean is actually greater than the mean itself. Moreover, we lack long, high-resolution proxy records of past monsoon intensity that could tell us about the monsoon's natural patterns. Recent research has focused on speleothem records of hydroclimate variability in the Southwest [5, 6, 7]. These records provide insight into the natural range of climate and water variability on decadal to millennial timescales.

The $\delta^{18}O$ value of cave calcite responds to climate on seasonal to millennial timescales [8, 9, 10]. However, a 1‰ change in precipitation $\delta^{18}O$ does not translate to an instantaneous 1‰ change in dripwater or calcite $\delta^{18}O$: dripwater $\delta^{18}O$ variability is damped compared to surface precipitation [6]. This is likely due to seasonal effects of temperature, evaporation, and runoff at the surface, subsurface storage and mixing of water, and variations in cave air pCO_2 and temperature [11, 12, 13, 14]. These myriad processes impart variability on cave dripwater that is not well quantified. Here, we develop a simple forward model to elucidate the impact of surface processes and subsurface storage on dripwater/calcite $\delta^{18}O$. This approach allows us to distinguish the relative impact of climate variability versus variability imparted by water storage in the soil and epikarst on calcite $\delta^{18}O$ values.

2. Data & Methods

2.1. Model

We develop a conceptual model in which precipitation falls at the surface with a characteristic $\delta^{18}O$ value, infiltrates into the soil, mixes with water already present in the subsurface, and drips into the cave, ultimately forming calcite. We represent these processes with a two-layer model: "soil" and "rock". In the soil layer we simulate soil moisture and soil water $\delta^{18}O$ by modifying the "leaky bucket" model [15] to conserve mass. This model uses soil porosity of ~47%, or a maximum water holding capacity of 760mm for a standard 1.6m column of soil. This approach uses the Thornthwaite equation for evaporation [16] and empirical soil parameter estimates from Huang et al. [15]. We do not explicitly resolve fractionation from evaporation because additional model experiments indicated that when fractionation matters most (summer), evaporation removes almost all the water before it enters the subsurface. Hence, integration of the soil layer requires only monthly precipitation amount, precipitation $\delta^{18}O$ and temperature as inputs. Subsurface flow from the soil layer is proportional to the moisture content of the layer. Its corresponding $\delta^{18}O$ value is then passed to the rock layer, which we represent as a large, well-mixed reservoir. We assume that moisture flux in the rock layer is in steady state equilibrium; outflow volume equals inflow volume at each time step. The simulated cave dripwater $\delta^{18}O$, therefore, is the $\delta^{18}O$ of the water leaving the rock layer. We assume this signal is passed directly to the speleothem during calcification because at a number of monitored caves in the Southwest, calcite precipitation appears to be in equilibrium, and thus calcite $\delta^{18}O$ variance is comparable to dripwater $\delta^{18}O$ variance. The model is not intended to reconstruct the time evolution of observed dripwater $\delta^{18}O$ values because the site-specific input time series of $\delta^{18}O$ is rarely known. Instead, our model captures important statistical characteristics of the dripwaters.

2.2. Data

We obtained monthly temperature and precipitation inputs for our model from the University of East Anglia Climate Research Unit (CRU) TS2.1 precipitation [17] and temperature [18] datasets. Values are derived from station data interpolated to a 0.5° grid; we used the grid point corresponding to Cave of the Bells (31° 45' N; 110°45'W). We acquired precipitation isotope values from a dataset produced by Eastoe and Long [19] in which Tucson precipitation amount, $\delta^{18}O$, and δD have been measured by event since 1981 [20]. We amount-weighted the isotope data by month. Winter precipitation averages ~195mm (ONDJFM) and has an average amount-weighted isotopic $\delta^{18}O$ value of ~ -10‰. Summer precipitation averages ~195mm (JAS) and has an average amount-weighted isotopic $\delta^{18}O$ value of ~ -6.5‰. We compare our model output with modern dripwater data collected since 2003 from Cave of the Bells (COB) and an existing 3000-year Holocene speleothem record from the same cave [21, 22].

2.3. Experimental Design

We drive our forward model with three sets of experiments using observational data, idealized climatologies, and stochastic input. We first compute a monthly climatology of precipitation amount, temperature, and precipitation $\delta^{18}O$ from the observational data. We initialize the model by driving it with this climatology for 300 years. The mean dripwater $\delta^{18}O$ output from this simulation is used as the "control mean". We then use the existing time series of precipitation $\delta^{18}O$, precipitation amount, and temperature, to drive our model for the last few decades. This produces one realization of dripwater $\delta^{18}O$ that could be the output of modern climate, the mean of which is used for comparison to other model experiments. These subsequent sensitivity tests were designed not to represent plausible climatologies (though some do) but rather to test the model response to various input parameters on short and long timescales.

In the first set of sensitivity tests, we isolate the role of seasonal evaporation by holding precipitation $\delta^{18}O$ and amount to climatology while separately varying summer and winter mean temperature values from climatology by ±4°C. Second, we maintain temperature and $\delta^{18}O$ at climatology and scale precipitation by 0% (no precipitation), 50%, 150% and 200% (doubling of precipitation) for each season (JFM, AMJ, JAS, OND). Third, we hold temperature and $\delta^{18}O$ at climatology, but change the length of the monsoon. That is, we substitute the July precipitation mean amount for June (i.e., early/long monsoon, as June is generally dry) and likewise substitute the June precipitation mean for July (i.e., late/short monsoon). In each case, we ran the model for 300 years after initialization and measured the response of the model to the given experiment as the difference in per mil from the "control mean". We do not vary precipitation $\delta^{18}O$ input in this study, although clearly changes in this variable will affect dripwater output.

The second set of experiments tests the effect of subsurface storage and mixing on dripwater $\delta^{18}O$. We used the observed range and mean of precipitation amount, temperature, and $\delta^{18}O$ to create synthetic time series with the same mean and variance as the observational data. To generate these series we first estimated the mean and variance of temperature and $\delta^{18}O$ observations for each month, then used these parameters to generate normally distributed data with the same statistical characteristics. A different approach was necessary for generating precipitation amounts, as we found monthly precipitation to be gamma-distributed: smaller rain events are more frequent than larger, giving a probability distribution skewed to smaller values. We therefore calculated maximum likelihood estimates of the size and shape parameters (k and θ) of the gamma distribution for each month [23]. From this, we generated random gamma-distributed precipitation estimates to use as model input. The synthetic time series have no temporal structure except the seasonal cycle, so that on interannual timescales their spectra are "white". Using these synthetic temperature, precipitation $\delta^{18}O$, and amount distributions, we designed a Monte Carlo experiment to test model sensitivity to changes in the reservoir size ratio (rock: soil reservoirs). This approach is necessary because using the 30-yr observational dataset to drive the reservoir size ratio tests might impart interannual or decadal variability that is not due solely to reservoir effects or subsurface storage. We vary the reservoir sizes from 5:1 to 1:1 and even 1:2, which is unrealistic in the case of COB where the soil is thin relative to the thickness of rock. The Monte Carlo simulations were run for 3000 years to be able to compare the results to oxygen isotope variability in the existing 3000-yr COB speleothem record. We analyze spectra of these synthetic drip time series [24] to assess the influence of subsurface processes at specific timescales.

3. Results

The means of the model $\delta^{18}O$ time series are comparable to the observed data from Cave of the Bells (figure 1a). Our model reproduces the observed dominance of winter precipitation on dripwater $\delta^{18}O$. The variance is also similar between modeled and observed dripwater; figure 1b shows the means and distributions of the observed cave drips, the modeled drips of varying size ratios, and input precipitation $\delta^{18}O$. Figure 1c compares observed and modeled dripwater $\delta^{18}O$ data. The observed damping effect from the surface to the cave is reproduced in the model dripwater data.

The mean modeled dripwater is close to and stable at the winter precipitation mean but can be modified by seasonal variability in precipitation amount and timing. The strongest/easiest way to modify model dripwater $\delta^{18}O$ mean is to turn off precipitation for a season (table 1). This is true for any season, but can be seen most dramatically in the absent or weakened monsoon and fall season precipitation. The timing of monsoon precipitation also matters in the model: doubling monsoon precipitation has a comparable effect on dripwater to a longer monsoon (+0.33‰ vs. +0.35‰). In the observational data, we see a large (+1‰) change in the $\delta^{18}O$ value of one drip only following the strong and long 2006 monsoon, suggesting that some drips may be more responsive to monsoon moisture than others.

The variance of dripwater $\delta^{18}O$ can also be modified, primarily by changing the size of the subsurface reservoirs feeding the drips. The model simulates an appropriate reduction in $\delta^{18}O$ variance. The observed ratio of precipitation $\delta^{18}O$ variance to dripwater $\delta^{18}O$ variance at COB is between 4 and 8%. For rock reservoir to soil reservoir ratios of 1:2, 1:1, and 5:1, model dripwater exhibits 12%, 5%, and 3% of the variance of precipitation $\delta^{18}O$ input. Thus, the larger the volume of the rock reservoir, the more damping of precipitation $\delta^{18}O$ occurs.

A comparison of the median variance spectrum from 100 Monte Carlo time series and an actual speleothem record from COB [21, 22] is shown in figure 2. At shorter timescales (<100 years), the Monte Carlo spectra are all somewhat reddened. At timescales greater than 100 years, however, these spectra all begin to resemble the synthetic input precipitation $\delta^{18}O$, which is flat on interannual timescales. The COB spectrum has a similar slope as the Monte Carlo spectra, but at timescales longer than 100 years, the COB spectrum exhibits higher variance than any of the modeled spectra.

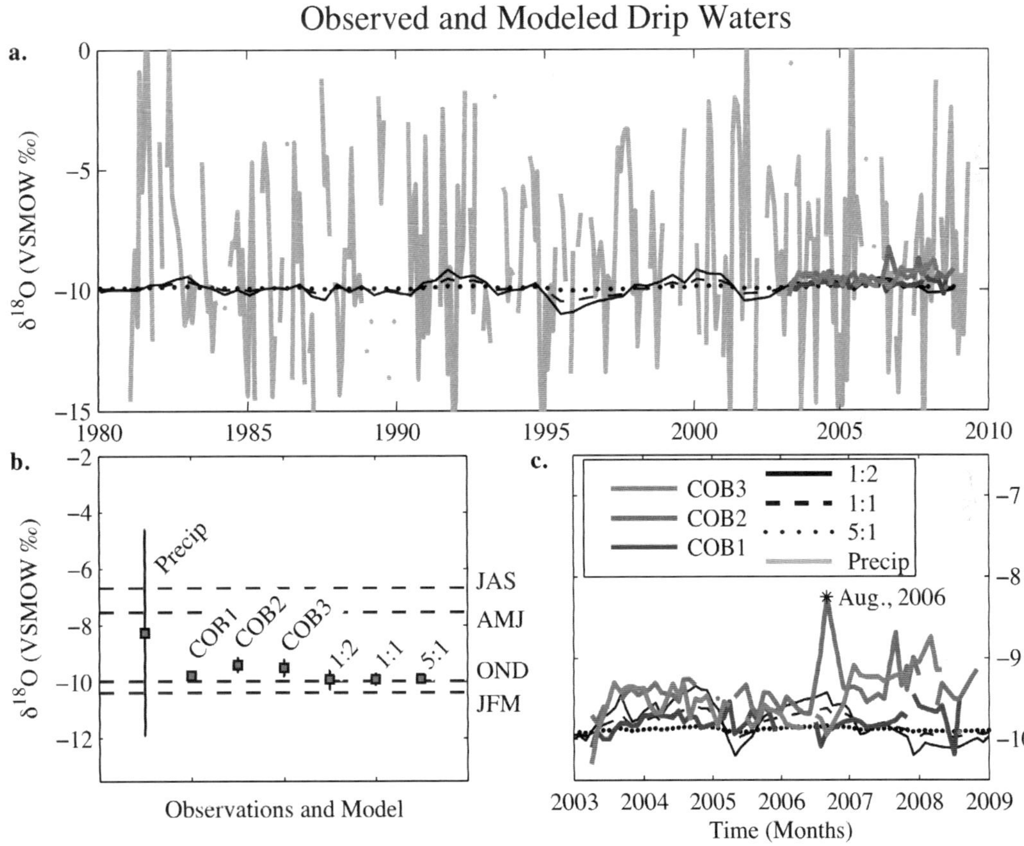

Figure 1. (a) Surface precipitation $\delta^{18}O$ (grey), modeled (black) and observed (colors) dripwater $\delta^{18}O$. Subsurface storage and mixing damp the surface precipitation signal in the observed system and the model. (b) Precipitation (model input), resultant dripwater $\delta^{18}O$ for rock: soil reservoirs 1:2, 1:1, and 5:1, and observed COB drips. Error bars represent 1 standard deviation. Note the same damping effect; changing the rock: soil moisture ratio affects the variance of the modeled drips. JAS = observed $\delta^{18}O$ mean for July-August-September; OND = October-November-December; JFM = January-February-March, and AMJ = April-May-June. (c) Observed (colors) and modeled (black) dripwater $\delta^{18}O$ values from Cave of the Bells. August 2006 is starred because this datapoint may represent the inflow of monsoon rain during and just after the strong monsoon of the summer of 2006. Note, however, that the different drips respond differently, and lag, the strong monsoon rainfall.

Table 1. Sensitivity test results for the forward model. Dripwater $\delta^{18}O$ responses are relative to the "control" run (precip x 1.0). Note that we did not alter the oxygen isotopic value of input precipitation.

Variable	Test	Season	Dripwater $\delta^{18}O$ response (‰)
Temperature	Warm (+4°C)	Summer	-0.03
	Cool (-4°C)		+0.07
	Warm (+4°C)	Winter	+0.04
	Cool (-4°C)		-0.02
Precipitation amount	Long Monsoon	Summer	+0.35
	Short Monsoon		-0.27
	precip x 0 (shutoff)	JFM	-0.58
		AMJ	-0.01
		JAS	-0.75
		OND	+0.62
	precip x 0.5	JFM	+0.04
		AMJ	-0.01
		JAS	-0.08
		OND	+0.25
	precip x 1.5	JFM	-0.03
		AMJ	-0.22
		JAS	-0.10
		OND	-0.08
	precip x 2 (doubling)	JFM	-0.05
		AMJ	-0.22
		JAS	+0.33
		OND	-0.15

4. Discussion

Our results indicate that mean dripwater $\delta^{18}O$ can be altered by changes in climatic inputs, but the variance exhibited by the dripwater $\delta^{18}O$ time series could be due in part to subsurface storage of water. We did not reproduce the temporal evolution of the observed COB dripwaters, but we did not expect to; we use isotope data from nearby Tucson, AZ, and given the spatial heterogeneity of $\delta^{18}O$ in precipitation in the arid Southwest, it is unlikely we would have captured specific elements of the time series of COB dripwaters. However, we modeled physically plausible drips that demonstrate characteristics consistent with observed dripwaters, which suggest that our results are also physically plausible.

The mean $\delta^{18}O$ of dripwater is generally similar to the winter precipitation average, but is modulated by monsoon length and/or total amount. The timing of precipitation, especially monsoon rainfall, is a particularly important control on the seasonal dependence of dripwater $\delta^{18}O$. Wetter soils early in the summer season (i.e., in June, when soils are very dry) allow more water to transit to the cave. This effect is more dramatic than doubling the amount of monsoon precipitation. Fall season precipitation is likely important for the same reason, as much of fall precipitation moistens soils and could push moisture through to the cave environment. Temperature changes do not have much of an effect, perhaps because ±4°C is insufficient to alter the precipitation minus evaporation budget substantially. In all of the experiments performed, we do not attain a dripwater $\delta^{18}O$ mean that is near the annual average of precipitation $\delta^{18}O$ (-8.2‰). Even when monsoon precipitation is doubled, the monsoon is underrepresented in dripwater. However, different types of flow pathways through limestone (e.g., fracture or fissure flow) could let more monsoon rain into the cave; this type of flow was not explicitly dealt with in our model.

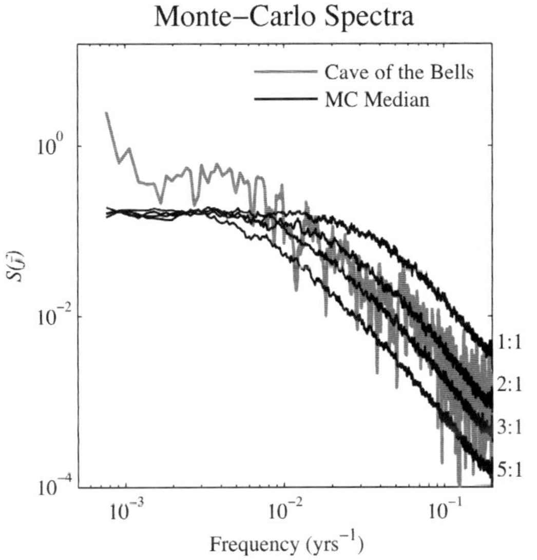

Figure 2. Median spectral densities from 100 Monte-Carlo simulations with different rock: soil reservoir ratios (black lines) and Cave of the Bells Holocene speleothem record (red line). Note that subsurface storage and mixing processes impart a certain level of background variability and redness to the dripwater $\delta^{18}O$ spectrum. On timescales longer than 100 years, soil processes can no longer be implicated in producing some of the variance in the COB speleothem record.

The most important variable controlling dripwater $\delta^{18}O$ variance is the ratio of reservoir sizes feeding the cave drips. This makes sense because the larger the moisture pool, the more damping there will be of a noisy surface precipitation signal, (e.g., Hasselmann [25]). This also means that short-term (decadal to multi-decadal) variability in surface precipitation cannot be linearly translated to cave dripwater. Determining actual moisture pools for real caves is a significant challenge. Monitoring is important to determine how much of surface precipitation variability, and what season of precipitation, is represented in the cave environment.

Spectral analysis of the COB speleothem and the Monte Carlo drip data shows that at longer (>100 year) timescales, $\delta^{18}O$ variability at Cave of the Bells is much stronger than any background variance imparted by soil/rock processes. However, at shorter timescales, the observed speleothem variance does not rise above that which could potentially be generated by subsurface mixing. This suggests that variability in this range can represent a mixed signal of climate input and subsurface storage. This effect is likely drip- and cave-dependent; it may also vary as background climate changes. Thus, subsurface moisture storage must be considered in paleoclimatic interpretations of speleothem data, particularly on sub-century timescales and in areas with a pronounced summer soil moisture deficit.

Observed dripwaters from Cave of the Bells show an almost exclusively winter rainfall $\delta^{18}O$ signal, except during years with exceptionally strong summer monsoons, which is consistent with our model results. Other dripwater studies suggest controls on different drips vary within and between caves [11, 26, 27, 28, 29]. If two drips within one cave are fed by completely different flow pathways, the same climate signal could conceivably produce very different speleothem $\delta^{18}O$ records. Clearly this indicates the need for paleorecord replication within and between caves. A strong correlation among records from different caves and drips will strengthen the case that calcite isotopic variability is more representative of precipitation $\delta^{18}O$ variability, and thus climate, than of subsurface processes.

5. Conclusions

Our oxygen isotope model of cave dripwater suggests that winter precipitation is the dominant contributor to the dripwater $\delta^{18}O$ mean, likely because the monsoon signal is removed by runoff and evaporation. However, extended/early monsoon conditions or doubling monsoon precipitation allows infiltration of some monsoon moisture. Altering fall season precipitation also impacts dripwater $\delta^{18}O$, perhaps disproportionately to the fall precipitation amount. Finally, the model suggests that subsurface storage and mixing, coupled with temporally uncorrelated climate input, can generate decadal variance. Observed variance in a COB speleothem at century to millennial scales, however, is greater than that expected from our model and thus likely reflects climate processes primarily. Constraints on residence time and subsurface moisture volume are needed to separate the relative influences of climate and storage on speleothem records on sub-century timescales. In a single record, subsurface storage can make interpretation of superimposed climate signals difficult. Such processes may not be consistent between caves or even necessarily between drips; consequently, replication offers a powerful strategy to separate the climatic signal from nonclimatic noise. This work demonstrates the importance of process-based studies in quantifying how paleoclimate proxy records filter and respond to climate.

Acknowledgements
We thank Christoph Spötl and an anonymous reviewer for comments on the manuscript. This work was supported by a Science Foundation Arizona Graduate Fellowship (SAT), a National Defense Science and Engineering Graduate Research Fellowship (SAT), a National Science Foundation Graduate Research Fellowship (TRA), and the National Science Foundation's ESH and P2C2 programs (JEC). Support to attend the PAGES Young Scientist's Meeting in July 2009 was provided by the National Science Foundation (SAT).

References

[1] Adams DK and Comrie AC 1997 *Bull. Amer. Met. Soc.* **78** 2197-2213
[2] Higgins RW, Yao Y, Wang XL 1997 *J. Climate* **10** 2600-22
[3] Seager R 2007 *J. Clim.* **20** 5527-5552
[4] Christensen JH et al. 2007 Regional Climate Projections. In: *Climate Change 2007: The Physical Science Basis. Contribution of Working Group I to the Fourth Assessment Report of the Intergovernmental Panel on Climate Change* [Solomon, S., D. Qin, M. Manning, Z. Chen, M. Marquis, K.B. Avery, M. Tignor and H.L. Miller (eds.)] (Cambridge: Cambridge University Press)
[5] Asmerom Y, Polyak V, Burns S and Rassmussen J 2007 *Geology* **35** 1-4
[6] Polyak VJ and Asmeron Y 2001 *Science* **294** 148-51
[7] Wagner JDM, Cole JE, Beck JW, Patchett PJ, Henderson GM and Barnett HR 2010 *Nat. Geosci.* **3** 110-113
[8] Johnson KR, Hu C, Belshaw NS, Henderson GM 2006 Earth and Planetary Science Letters **244** 394-407
[9] Genty D, Baker A, and Vokal B 2001 *Chemical Geology* **176** 191-212
[10] Wang YJ, Cheng H, Edwards RL, An ZS, Wu JY, Shen C-C, and Dorale JA 2001 *Science* **294** 2345-8
[11] Lachniet MS 2009 *Quat. Sci. Rev.* **28** 412-32
[12] Mattey D, Lowry D, and Duffet J 2008 *Earth and Planetary Science Letters* **269** 80-95
[13] Banner J, Guilfoyle A, James EW, Stern LA, and Musgrove M 2007 *J. Sedimentary Research* **77** 615-22
[14] Baker A, and Brunsdon C 2003 *J. Hydrology* **277** 151-63
[15] Huang J, Van Den Dool HM and Georgarakos KP 1996 *J. Clim.* **9** 1350-62
[16] Thornthwaite CW 1948 *Geogr. Rev.* **38** 55-94
[17] Mitchell TD and Jones, PD 2005 *Int. J. Climatology* **25** 693-712
[18] Brohan P, Kennedy JJ, Harris I, Tett SFB and Jones PD 2006 *J. Geophys. Res.* **111** D12106
[19] Eastoe, C. and Long, A. Unpublished data on precipitation isotopes in Tucson, AZ
[20] Wright WE, Long A, Comrie A, Leavitt SW, Cavazos T, Eastoe C 2001 Geophys. Res. Lett. **28** 787-90
[21] Wagner JDM 2006 UA Dissertation 129pp
[22] Cole et al. in prep.

[23] Hahn GJ and Shapiro SS 1994 *Statistical Models in Engineering* (New York: John Wiley & Sons) p 88
[24] Thomson DJ 1982 Proceedings of the IEEE **70** 1055-96
[25] Hasselmann K 1976 Tellus, Series A, Dynamic meteorology and oceanography **28** 473–85
[26] Ayalon A, Bar-Matthews M, Sass E 1998 *J. Hydrology* **207** 18–31
[27] Cobb KM, Adkins JF, Partin JW, Clark B 2007 *Earth and Planetary Science Letters* **263** 207–20
[28] Baldini JUL, McDermott F, and Fairchild IJ 2006 *Chemical Geology* **235** 390-404
[29] Tooth AF and Fairchild IJ 2003 *J. Hydrology* **273** 51-68

IOP Conf. Series: Earth and Environmental Science **9** (2010) 012023 doi:10.1088/1755-1315/9/1/012023

Evaluation of the "amount effect" at speleothem sites in the Asian monsoon region

Jung-Eun Lee[1] and Abigail L Swann[2]

1. Department of Geophysical Sciences, The University of Chicago, Chicago, IL 60637

2. Department of Earth and Planetary Science, University of California, Berkeley, CA 94720

jelee@uchicago.edu

Abstract. Oxygen isotope measurements from speleothems have been used to infer past hydrological cycles because the isotopic composition of precipitation is observed to decrease with increasing precipitation rates over low latitude coastal and island stations (the "amount effect"). Here we show that local precipitation amount can be inferred from the differences in oxygen isotopic composition if the measurement site is in a subtropical coastal region where water vapor is transported directly from the ocean. The amount effect works because vapor from local evaporation and vapor imported from other regions have distinctly different isotopic compositions—the $\delta^{18}O$ of vapor flux from evaporation is ~10‰ higher than the $\delta^{18}O$ of imported vapor over subtropical oceanic regions—and precipitation is usually higher over the area with larger convergence. Thus, $\delta^{18}O$ in precipitation can differentiate between the contribution of local evaporation and moisture convergence if the isotopic composition of local evaporation and or transported vapor remains relatively constant. Our work indicates that sites further inland may not be suitable for studying the changes in local precipitation amount but still reflect the combined changes of precipitation amount and changes in vapor transport pattern. We have shown that Hulu Cave is a reasonable site to interpret precipitation from $\delta^{18}O_p$: estimated precipitation difference between present-day and Last Glacial Maximum (LGM) from measured $\delta^{18}O_p$ difference (1~2‰) would be 0.8-2.2 mm/day (lower during the LGM) assuming our model simulates evapotranspiration reasonably well at this site.

1. Introduction

The oxygen isotopic composition of speleothem calcite has been widely used to reconstruct past hydrological variations because the $\delta^{18}O$ of speleothem calcite has been shown to record $\delta^{18}O$ of precipitation ($\delta^{18}O_p$; 1). Observed isotopic composition of precipitation ($\delta^{18}O_p$) is inversely correlated with precipitation amount over low latitude island and coastal regions [2, 3]. This relationship is called the "amount effect" and relates the $\delta^{18}O_p$ and precipitation amount at different locations for present-day climate. This relationship is frequently applied to estimate precipitation in the past at a fixed location. However, it has been challenged by other paleo-wetness proxies [4], and [5] and [6] point out that the $\delta^{18}O_p$ in the Asian monsoon region reflects hydrological processes upstream and not primarily the amount of precipitation at the site.

Mean atmospheric water vapor has a lower $\delta^{18}O$ than the evaporated vapor because atmospheric vapor has been isotopically depleted by condensation and subsequent interaction with falling raindrops after being evaporated from the ocean [7]; the isotopic composition of evaporated vapor is ~10‰ higher than that of near-surface atmospheric vapor over subtropical oceanic regions. Regions with high rates of moisture convergence generally also have high rates of precipitation. In these high convergence regions, $\delta^{18}O_p$ has a relatively larger contribution from transported atmospheric vapor compared to local evaporation since rates of evaporation tend to be less variable than precipitation. $\delta^{18}O_p$ can differentiate between the contribution from local evaporation and moisture convergence if the isotopic composition of vapor from local evaporation and transported vapor remains relatively constant. If these conditions are met, the amount effect holds and local measurements of $\delta^{18}O_p$ can be used to infer local precipitation. This concept was discussed in [8] relating local precipitation amount and the $\delta^{18}O_p$ of precipitation over South America. On a storm scale, diffusive processes that act while raindrops are falling and the subsidence of cold and isotopically depleted vapor decrease $\delta^{18}O$ in vapor and precipitation [9, 10, 11]

To evaluate the usage of stable isotopes in interpreting past changes in the hydrological cycle, we use simulation results from an isotope-enabled climate model. We focus on the Asian monsoon region because many speleothem records from Asian caves have been used to interpret glacial-interglacial changes in hydrological cycle in Asia [12, 13, 14, 15].

2. Model description

Water isotopes are incorporated into the National Center for Atmospheric Research (NCAR) Community Atmospheric Model Version 2 (CAM2) [7], and have been used to evaluate the usage of stable isotopes as proxies for past temperatures and past precipitation amounts. We integrated the isotope-enabled NCAR CAM2 with fixed sea surface temperature (SST) and sea ice distributions for the present-day (PRS) (observed) and the Last Glacial Maximum (LGM) (CCSM results; 16). SST in tropics in tropics (20°S-30°N) is lower than present-day by ~2.6°C. Detailed information on the model simulations can be found in [17].

We focus here on the Asian monsoon region including the Indian and eastern Asian monsoon regions. The NCAR CAM2 overestimates precipitation over the Indian monsoon region and underestimates precipitation over the East Asian monsoon region (Figure 2 of ref. 7). This is possibly due to the poor simulation of topography of the Tibetan Plateau, which effects the precipitation simulation during the present-day and the LGM. Despite the poor representation of precipitation, this work can be thought of as a test of the hypothesis that $\delta^{18}O_p$ can represent local precipitation amount as many analyses of speleothem measurements assume.

3. Results and Discussion

The ideal situation to test the hypothesis that $\delta^{18}O_p$ represents local precipitation is to measure precipitation and $\delta^{18}O_p$ at each point of interest at different times and compare the relationship. Figure 1 shows such a map from model simulations—precipitation and $\delta^{18}O_p$ difference between the present-day and the LGM. If $\delta^{18}O_p$ is an indicator for precipitation amount, the precipitation and $\delta^{18}O_p$ should have opposite sign, or the same colors in Figure 1 (color scales for precipitation and $\delta^{18}O_p$ difference are opposite). For most of the subtropical coastal region (20°N-30°N), the relationship seems to hold as [8] indicates.

The most notable feature in Figure 1b is a large area of higher $\delta^{18}O_p$ (west of 100°E). The model simulates a large decrease in precipitation in the Indian monsoon region during the LGM, and the $\delta^{18}O_p$ is higher over most of the surrounding regions—even over regions where an increase in precipitation is simulated (for example, regions around 90°E-100°E and 20°N-30°N and over the Bay of Bengal). This is due to the change in moisture source region from the east (present-day) to west

(LGM): most of the water vapor from the east (around Indonesia) is much more depleted due to the large amount of precipitation there for both the present-day and the LGM. As a result, $\delta^{18}O_p$ during the LGM is higher even where the precipitation amount is larger than present-day.

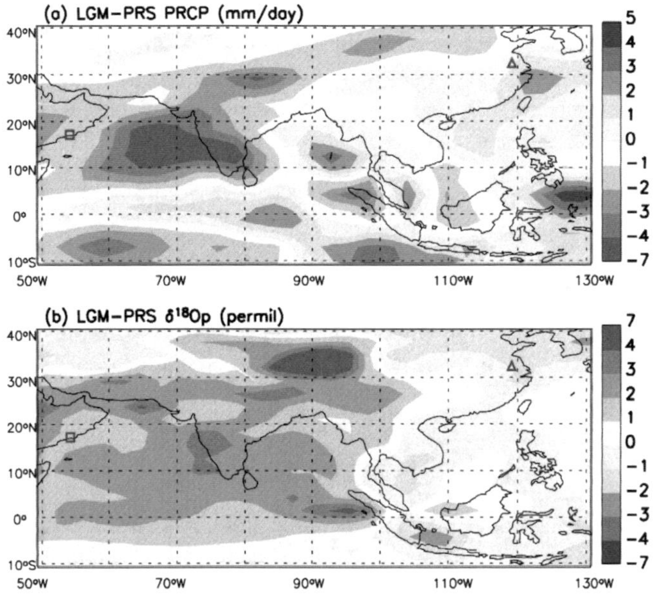

Figure 1. The LGM-present differences in (a) precipitation and (b) $\delta^{18}O_p$. The triangle and square denote the location of the Hulu Cave [12] and Qunf Cave [13] respectively.

The areal extent of the $\delta^{18}O$ signature in vapor from changes in precipitation may be much larger than the region in which precipitation changes. In this way, large changes in remote precipitation can modify the local $\delta^{18}O_p$ by changing the isotopic values of incoming vapor from advection. The decrease in precipitation over India in LGM (Figure 1a) causes a large areal change in $\delta^{18}O_v$ (Figure 2c), which is advected to the downstream regions and influences the local $\delta^{18}O_p$(big red blob of higher $\delta^{18}O_p$ around 95°E, 33°N in Figure 1b and 2c). The isotopic values of speleothem records may not represent precisely local changes in precipitation, but are instead showing an integral of the precipitation changes over a larger area.

The locations of two notable speleothem records are marked in Figure 1: the Hulu cave (32°30′N, 119°10′E; square in Figure 2c) in China and the Qunf Cave (17°10′N, 54°18′E, cross in Figure 2c) in Oman. The model simulated $\delta^{18}O_p$ difference from the LGM to the present-day in Oman is about -4‰ (Table 1). As far as we are aware, speleothem values from the LGM are not published for this cave, possibly because there was no calcite growth due to dry conditions there. The LGM value of $\delta^{18}O_p$ estimated from speleothems in the Hulu Cave is higher than present-day values by 1-2‰ [18], which is the opposite sign compared to the model $\delta^{18}O_p$ difference (about -0.4‰). We suspect that this discrepancy is due to the model deficiency in simulating the Asian monsoon.

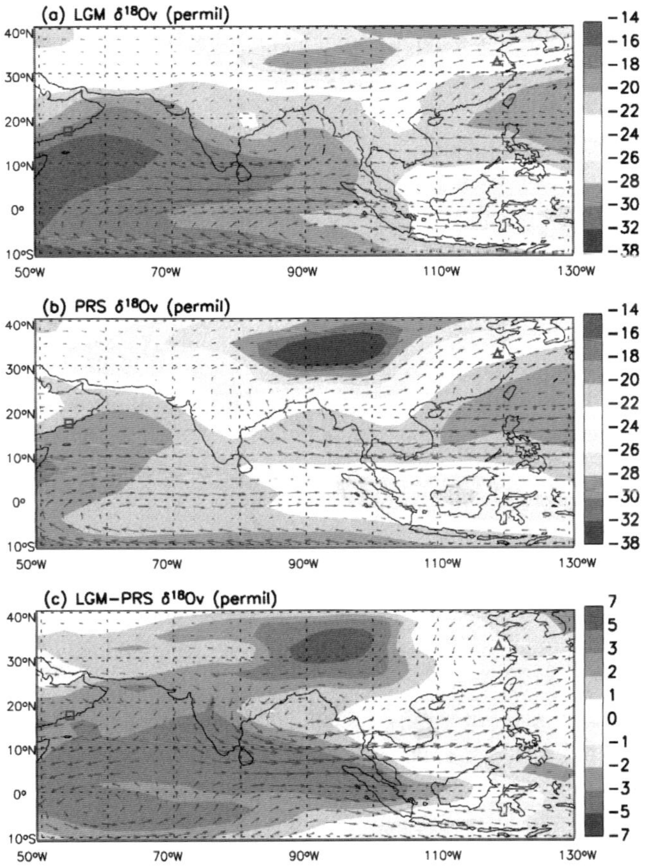

Figure 2. Isotopic composition of the annual mean column vapor for (a) LGM, (b) PRS and (c) the difference. Arrows are annual mean total column water vapor transport.

Table. 1. Precipitation (P; mm/day) and evaporation (E; mm/day) in Qunf cave (17°10′N, 54°18′E, square in Figure 1) and Hulu Cave (32°30′N, 119°10′E; triangle in Figure 1) together with the isotopic composition of P and E over this region. Mean isotopic composition (‰) of each component *i* is represented as $\delta^{18}O_i$. Estimated values are from the modified Dansgaard slope (see text).

	Qunf Cave				Hulu Cave			
	PRS	LGM	LGM-PRS		PRS	LGM	LGM-PRS	
			Modeled	Estimated			Modeled	Estimated
P	2.35	0.42	-1.93	-6~-4.4	2.86	3.30	0.44	~0.43
E	4.05	3.10	-0.95		2.62	2.34	-0.28	
$\delta^{18}O_p$	-3.96	-1.08	2.88		-6.11	-6.50	-0.39	
$\delta^{18}O_e$	-6.37	-7.49	-1.12		-6.19	-7.24	-1.05	

The present-day slope of precipitation amount vs. $\delta^{18}O_p$ ranges from -0.48 ‰/(mm/day) to -0.66 ‰/(mm/day) for some tropical/subtropical island stations [2]. Using the present-day slope (Dansgaard slope), we can extrapolate to find changes in the LGM precipitation based on measured $\delta^{18}O$ records.

If the isotopic composition of the incoming vapor is constant in time, we expect this relationship to give us a good estimate of the amount of precipitation through time. At the Qunf cave, the differences in the model estimate of $\delta^{18}O_p$ of 2.88‰ can be interpreted, using the Dansgaard slope, as a 4.4-6 mm/day decrease in precipitation during the LGM—a much larger change than the modeled decrease of 1.9 mm/day (Table 1). Because the $\delta^{18}O_p$ of incoming vapor is higher during the LGM over this region (Figure 2), changes in the precipitation amount cannot fully explain the $\delta^{18}O_p$ change at the Qunf cave.

Over the tropical ocean, $\delta^{18}O_p$ can be approximated to vary with precipitation but when evaporation rates change the relationship must be scaled so that $\delta^{18}O_p$ varies as P/E (c.f., 7). Over the Hulu cave, the modeled evaporation rate is lower (~ 2.5 mm/day) than over the tropical oceanic regions (~4 mm/day). Using $\delta^{18}O_p$ scaled by P/E, the slope for $\delta^{18}O_p$ vs. precipitation at the Hulu Cave equals -0.77 to -1.1 ‰/(mm/day). The modeled change in $\delta^{18}O_p$ at the Hulu Cave is -0.39‰, which can be interpreted (using the modified Dansgaard slope) as an increase in precipitation by 0.35 mm/day to 0.50 mm/day. This agrees well with precipitation from model results that show an increase in precipitation of 0.44 mm/day, which is close to the estimates from the Dansgaard slope. The good agreement between the modeled precipitation rate and the rate calculated from modeled values with the modified Dansgaard slope indicates that the isotopic composition of incoming vapor is constant in time, which is verified in Figure 2.

4. Conclusions

Speleothem $\delta^{18}O$ measurements have been interpreted as indicators of local precipitation amount, and used to establish a link between the monsoon strength and temperature difference in Greenland [12, 13]. In order to infer local precipitation from measurements of $\delta^{18}O_p$, the isotopic composition of incoming vapor and local evpotranspiration must stay constant. Otherwise, measurements of $\delta^{18}O_p$ only reflect changes in upstream $\delta^{18}O_v$. On a glacial-interglacial time-scale, the incoming vapor in subtropical coastal regions exhibits a relatively small change in isotopic composition, making these regions suitable for studying precipitation change using $\delta^{18}O_p$ (c.f., 8). We have shown that Hulu Cave is a reasonable site to interpret precipitation from $\delta^{18}O_p$: estimated precipitation difference between present-day and Last Glacial Maximum (LGM) from measured $\delta^{18}O_p$ difference (1~2‰) would be 0.8-2.2 mm/day (lower during the LGM) assuming our model simulates evapotranspiration reasonably well at this site. Our work indicates that sites further inland than the Hulu Cave may not be suitable for studying the changes in local precipitation amount (Figure 1). These results show the conditions in only one model but this analysis still identifies which processes may be important and thus should be considered in the interpretation of any speolthem record.

Even if $\delta^{18}O_p$ cannot be used to infer local precipitation at some locations, it still contains useful information about the hydrological cycle in the past. The $\delta^{18}O$ variability in precipitation and atmospheric water vapor is largely driven by horizontal advection [20] and is shown to have a larger spatial footprint than precipitation [18]. Over regions where water vapor is advected from inland regions, the isotopic composition of transported vapor often reflects the precipitation changes in the upstream region. In this case, changes in $\delta^{18}O_p$ do not represent changes in the local precipitation amount, but changes in the hydrological cycle in the upstream region. If the change in the hydrological cycle is large, as in the case of the Indian monsoon region in our model, large upstream regions show similar changes in $\delta^{18}O_p$. Caution should be used in interpreting speleothem records as records of local precipitation amount alone, and the location of speleothem records should be considered when attributing changes in $\delta^{18}O_p$ to either moisture transport or precipitation. Thus, speleothem $\delta^{18}O$ records have great potential to provide information about past moisture transport and precipitation through investigating spatial patterns in past $\delta^{18}O_p$ even if they are unable to constrain local precipitation directly.

References

[1] Cobb, K. M., Adkins, J. F., Partin, J. W. and Clark, B. (2007), Regional-scale climate influences on temporal variations of rainwater and cave dripwater oxygen isotopes in northern Borneo. *Earth Plant. Sci. Lett.*, **263**, 207–220.

[2] Dansgaard, W. (1964), Stable isotopes in precipitation. *Tellus*, **16**, 436-468.

[3] Rozanski, K., Araguas-Araguas, L., and Gofiantini, R. (1993), Isotopic patterns in modern global precipitation, *Geoph. Monog.*, **78**, 1-36.

[4] Maher, B. A. (2008), Holocene variability of the East Asian summer monsoon from Chinese cave records. a reassessment. *The Holocene*, **18**, 861-866.

[5] Vuille, M., M. Werner, R. S. Bradley, and F. Keimig (2005), Stable isotopes in precipitation in the Asian monsoon region. *J. Geophys. Res.*, **110**, D23108, doi:10.1029/2005JD006022.

[6] LeGrande, A.N., and G.A. Schmidt (2009), Sources of Holocene variability of oxygen isotopes in paleoclimate archives. *Clim. Past*, **5**, 441-455. Lee, J.-E., I. Fung, D. J. DePaolo, and C. C. Henning (2007), Water isotopes during the Last Glacial Maximum: New general circulation model calculations, *J. Geophys. Res.*, **113**, D19109, doi:10.1029/2008JD009859.

[7] Lee, J.-E., I. Fung, D. J. DePaolo, and C. C. Henning (2007), Water isotopes during the Last Glacial Maximum: New general circulation model calculations, *J. Geophys. Res.*, **113**, D19109, doi:10.1029/2008JD009859.

[8] Lee, J.-E., K. Johnson, and I. Fung (2009), Tropical precipitation during the Last Glacial Maximum: An analysis of the "amount effect" with a water isotope-enabled General Circulation Model. *Geophys. Res. Lett.*, **36**, L19701, doi:10.1029/2009GL03925.

[9] Risi, C., S. Bony, F. Vimeux, M. Chong, and L. Descroix (2009), Evolution of the stable isotopic composition of the rain sampled along Sahelian Squall lines. *Q. J. R. Meteorol. Soc.*, doi: 10.1002/qj.485.

[10] Bony, S., C. Risi, and F. Vimeux (2008), Influence of convective processes on the isotopic composition (d18O and dD) of precipitation and water vapor in the tropics: 1. Radiative-convective equilibrium and Tropical Ocean – Global Atmosphere – Coupled Ocean-Atmosphere Response Experiment (TOGA-COARE) simulations, *J. Geophys. Res.*, **113**, D19305, doi:10.1029/2008JD009942.

[11] Lee J. E. and I. Fung (2008), Amount effect of water isotopes and quantitative analysis of post-condensation processes, *Hydrol. Process.*, doi: 10.1002/hyp.6637.

[12] Wang, Y. J., Cheng, H., Edwards, R. L., An, Z. S., Wu, J. Y., Shen, C. C., Dorale, J. A. (2001), A High-resolution absolute-dated late Pleistocene monsoon record from Hulu Cave, China. *Science*. **294**, 2345-2348.

[13] Fleitmann, D., Burns S. J., Mudelsee, M., Neff, U., Kramers, J., Mangini, A., and Matter, A. (2003), Holocene forcing of the Indian monsoon recorded in a stalagmite from Southern Oman, *Science*, **300**, 1737-1739.

[14] Yuan, D., H. Cheng, R. L. Edwards, C. A. Dykoski, M. J. Kelly, M. Zhang, J. Qing, Y. Lin, Y. Wang, J. Wu, et al. (2004) Timing, duration, and transitions of the last interglacial Asian monsoon. *Science*, **304**, 575-578.

[15] Cheng, H., R. L. Edwards, W. S. Broecker, G. H. Denton, X. G. Kong, Y. J. Wang, R. Zhang, and X. F. Wang (2009), Ice age terminations. *Science*, **326**, 248–252.

[16] Otto-Bliesner, B., E. Brady, G. Clauzet, R. Tomas, S. Levis, and Z. Kothavala (2006), Last Glacial Maximum and Holocene climate in CCSM3. *J. Clim.*, **19**, 2526-2544.

[17] Lee, J.-E., I. Fung, D. J. DePaolo, and B. Otto-Bliesner (2008), Analysis of the global distribution of water isotopes using the NCAR atmospheric general circulation model. *J. Geophys. Res.*, **112**, D16306, doi:10.1029/2006JD007657.

[18] Lee, J.-E., R. Pierrehumbert, A. Swann, and B.R. Lintner (2009), "Sensitivity of the stable water isotopic values on the convective parameterization schemes," *Geophys. Res. Lett.*, **36**, L23801, doi:10.1029/2009GL040880

[19] Wang, Y. J., Cheng, H., Edwards, R. L., Kong, X. G., Shao, X. H., Chen, S. T., Wu, J. Y.,

Jiang, X. Y., Wang, X. F., An, Z. S. (2008), Millennial- and orbital-scale changes in the East Asian monsoon over the past 224,000 years. Nature. 451, 1090-1093.

[20] Yoshimura, K., T. Oki, N. Ohte, and S. Kanae (2003), A quantitative analysis of short-term ^{18}O variability with a Rayleigh-type isotope circulation model. *J. Geophys. Res.*, **108**, 4647, doi:10.1029/2003JD003477.